WILDLIFE, WAR, AND GOD:

INSIGHTS ON SCIENCE AND GOVERNMENT

WILDLIFE, WAR, AND GOD:

INSIGHTS ON SCIENCE AND GOVERNMENT

MATTHEW A. CRONIN

LIBERTY HILL PUBLISHING

Liberty Hill Press
2301 Lucien Way #415
Maitland, FL 32751
407.339.4217
www.libertyhillpublishing.com

Printed in the United States of America.

LCCN: 2019911979

ISBN-13: 978-1-5456-7296-9

TABLE OF CONTENTS

ACKNOWLEDGEMENTS

Many friends and colleagues contributed to the ideas and insights in this book especially John Andre, Doyel Shamley, Bob Fanning, Bill Spearman, Todd Devlin, David Spady, Ted Lyon, Jim Beers, Lynn Noel, Shawn Haskell, Rob Ramey, Kent Holsinger, Dave Williams, Alan Maki, Ray Jakubczak, George Siter, Bill Streever, Valerius Geist, Ron Kahlenbeck, Bret Burroughs, Dave Harbour, Ernie Vyse, David Cameron, Thomas M. Cronin, Susan Crockford, Carol Lewis, Allen Mitchell, Jay McKendrick, Mike MacNeil, Tom Geary, Mark Peterson, Lance Vermeire, Robert Baker, Robert Bradley, Benny Gallaway, Mark Earnhardt, John Monarch, and many others including my co-authors in the Literature Cited.

Colonel Matthew Travis (USMC retired), Major Lucas Balke (USMC), Colonel Térèse D. LeFrançois (U.S. Air Force retired), and Colonel Paul Olsen (USA retired), provided very useful insights for Chapter 7. Dave Dilegge, Editor of *Small Wars Journal*, kindly gave permission to use my article (Cronin 2018). Kerry Halladay and Pete Crow provided insights and permission to use my articles in the *Western Livestock Journal*, and Scott Hollembaek generously provided information on wood bison in Alaska.

Special inspiration was provided by two deceased friends: Ben Hilaker, who was a U.S. Navy Corpsman in the Korean conflict and a fine man who understood wildlife management; and Warren Ballard who was an exemplary wildlife biologist.

My brothers Sean, Tom, Denis, Tony, Xavier, my sister Mary, my late parents Thomas and Corinne Cronin, and my children Colleen, Luke, and Jack provided support and motivation.

I owe special thanks to Addie Ruiz McEwen for encouragement that motivated me to write this book.

I also acknowledge the working men and women of the oil, timber, mining, and agriculture industries and past and present members of the U.S. Armed Forces who make our modern society possible.

I am responsible for information and ideas presented in the book, and all omissions and errors are mine.

This book is dedicated to the memory of Nikolai Vavilov

PREFACE

This book is about science and government policies on wildlife, the military, and religion. It is based on my education, training, and experience in the academic, government, and private sectors. It is not an up-to-date review and authority on each issue. Rather, I provide my experiences and perspectives as a scientist on the serious issues previously analyzed by others in books on the Endangered Species Act and other environmental issues in Chapters 1-6 (Chase 1987, 1995, Coffman 1994, Pombo and Farah 1996, Fitzsimmons 1999, Stirling 2008, Andre 2011, Lyon and Graves 2014, Crockford 2019) and on integration of the sexes in the military in Chapter 7 (Mitchell 1989, 1998, van Creveld 1991, 2001, Gutmann 2000, Browne 2007, Maginnis 2013). I am very impressed with the quality and thoroughness of these books, but they apparently have been ignored or dismissed by the government in making policies. If this book does nothing more than draw attention to these books, it will be a success.

The book is composed of Part 1 which is about conflicts in wildlife science, and Part 2 which describes general conflicts in science affecting society. Each chapter has a summary in bullet format (i.e., short statements), followed by more detailed information, and for two chapters an Appendix. Scientific Latin names of the species mentioned in the book are also in an Appendix. I cite representative literature which can be consulted for additional references on the issues that I address.

The bullet summaries are combined at the beginning of the book for easy access and can be used as independent descriptions of the issues.

These summaries may be useful for elected officials, business executives, their staffs, and others who need brief descriptions of issues. I attempted to make this book understandable to a broad audience with varying interests in scientific detail. I hope it is useful to people in government, industry, academia, the news media, the interested public, and especially young people who will be dealing with increasingly difficult environmental issues in the future.

I've often been told that environmental issues are entirely political and that the science doesn't matter, so I shouldn't take biased science so seriously. But I believe the science does matter because politically-motivated bias in science has historically had serious negative impacts on people and society (see Chapter 9). I therefore feel that it is my duty to write this book because I've been blessed with a career in science, natural resources, and agriculture made possible by the prosperity and freedoms for which Americans have worked, fought, and died. The prosperity and freedom that supports American science is ultimately dependent on a strong economy, which in turn is dependent on a strong military to defend it. Sustainability is a pervasive buzzword today, used to advocate for environmental policies. However, it is too often taken for granted that True Sustainability is reliant on all aspects of our society, with the economy and the military being its foundation.

The emergence of admitted socialists/communists in the U.S., including elected officials, is the most important factor that motivated me to write this book. The example of Lysenkoism in the former Soviet Union (Chapter 9) provides a chilling view of the consequences of the misuse of science to advance a socialist/communist agenda. Government control of science and the association of socialism/communism with environmentalism (Chapters 1-6), feminism (Chapter 7), and atheism (Chapter 8) provides a unifying theme of the book.

I think of the Korean and Viet Nam conflicts as battles in the larger Cold War against socialism/communism. We won the Cold War, but now we must confront the threat of socialism/communism within the United States. I oppose socialism/communism, and it is my hope this book helps to ensure that the sacrifices of our Soldiers, Sailors, Airmen, Marines, and Coastguardsmen in these battles, and the Cold War as a whole, will not have been in vain.

Matthew A. Cronin
June, 2019
Bozeman, Montana

BULLET SUMMARIES FOR ALL CHAPTERS

Chapter 1 Introduction: Science, policy, and management

Bullet Summary
— The Endangered Species Act (ESA) is the most powerful environmental law in America.
— The federal government makes scientific determinations for the ESA.
— The only recourse for challenging ESA determinations is to sue in court.
— This is largely ineffective because the government agencies apparently have unlimited funds for legal challenges and receive deference in court.
— This results in a government monopoly of the science and information with the ESA.
— In my experience, ideas, information, and science that are not in agreement with the federal government bureaucracy are dismissed or ignored.
— Prioritizing wildlife conservation and selective use of science has resulted in a bias against natural resource use and agriculture in the government bureaucracy and its associates in academia and environmental groups.
— This is important because environmentalism has become a tool of socialists/communists trying to undermine the basic structure of the U.S. Constitutional Republic.
— Environmental regulation, especially the taking of private property and federal seizure of States' jurisdiction of fish and wildlife with the ESA violates the U.S. Constitution.

— Elected officials, military personnel including veterans, and law enforcement officers, have all sworn an oath to uphold and defend the Constitution.

— Elected officials allowing the ESA to violate the Constitution appear to be dishonoring this oath.

— Several books have documented government misuse of science and policy with the ESA and other regulations (Coffman 1994, Chase 1995, Pombo and Farah 1996, Fitzsimmons 1999, Stirling 2008, Andre 2011, Lyon and Graves 2014, Crockford 2019).

— In this book I provide scientific information and insights that will complement these books.

Chapter 2. The Endangered Species Act: What does Endangered mean?

Bullet Summary

— The Endangered Species Act (ESA) is the most powerful environmental law in America.
— The ESA definition of wildlife species includes species, subspecies, and distinct population segments (DPS).
— This expands the ESA beyond its original scope of preventing extinction of species.
— Species may be listed under the ESA as either endangered or threatened with extinction.
— An endangered species is a species which is in danger of extinction throughout all or a significant portion of its range.
— A threatened species is a species which is likely to become an endangered species (i.e., likely to become endangered with extinction) within the foreseeable future throughout all or a significant portion of its range.
— It is important to recognize that the ESA applies to species that are at risk of extinction.
— Extinction means there are none (zero) remaining.
— The ESA should not apply to species that are simply experiencing declines in number or habitat and not at risk of extinction.
— Determinations of whether species are endangered or threatened with extinction are not scientifically definitive and made by government agencies.
— What constitutes a significant portion of a species' range and what constitutes the foreseeable future are not scientifically definitive and are determined by government agencies.
— An example is the greater sage grouse that was found to be warranted for ESA listing in 2010.

— The greater sage grouse was not endangered with extinction but the ESA listing was justified with predictions and speculation regarding its future habitat and numbers.

— The greater sage grouse was reassessed and found not warranted for ESA listing in 2015.

— The ESA allows the government to regulate use of private property. A voluntary incentives program for landowners instead of government regulation with the ESA would be more effective in protecting wildlife and consistent with American property rights.

Chapter 3. The Endangered Species Act: What is a Species?

Bullet Summary
— The Endangered Species Act (ESA) definition of "species" includes species, subspecies, and distinct population segments.
— This greatly expands the scope of the ESA beyond preventing extinction of species.
— There are several scientific definitions of species, but species are generally recognized as interbreeding populations that cannot successfully interbreed with other species.
— Wildlife species are generally identifiable, although some species designations are uncertain.
— Subspecies and distinct population segments (DPS) are populations within a species, that is, a part of a species, not the entire species.
— Subspecies are populations in a geographic area that differ from populations in other areas in traits such as size, color, and genetics.
— Subspecies are not scientifically definitive and subspecies designations are often subjective.
— Subspecies determinations for the ESA are made by government agencies.
— Distinct population segments (DPS) are populations designated considering their discreteness and significance.
— Discreteness refers to a DPS's differences from other populations. It is similar to the criteria that distinguish subspecies. DPS can also be separated by international borders.
— Significance refers to the importance of a DPS to the entire species and if it is unique in various ways.
— Discreteness and significance are not scientifically definitive.
— Discreteness and significance determinations for the ESA are made by government agencies.
— DPS are not scientifically definitive.

— DPS determinations for the ESA are made by government agencies.
— DPS for Pacific salmon have been named evolutionarily significant units (ESU).
— ESU determinations are not scientifically definitive.
— ESU determinations for the ESA are made by government agencies.

Chapter 4. Polar bears, Climate Change, and the Endangered Species Act

Bullet Summary

— Polar bears were listed as a threatened species likely to become endangered with extinction under the Endangered Species Act in 2008.

— The entire polar bear species was considered likely to become threatened with extinction because of loss of sea ice in the Arctic caused by global warming and climate change.

— This was determined with models predicting declines of sea ice and its impact on polar bear populations.

— I reviewed the determination that polar bears were a threatened species by the U.S. Fish and Wildlife Service and supporting research by the U.S. Geological Survey.

— I found that the determination of polar bears as threatened with extinction was premature and not scientifically rigorous.

— In my review I suggested that the prediction of a decline of polar bear numbers should be formulated as a hypothesis to be tested with observations.

— In my review I noted that polar bears survived previous warm periods.

— Subsequent published research suggests that polar bears and other species associated with sea ice survived previous warm periods in which there was likely little or no Arctic summer sea ice.

— This suggests they could also survive ice-free summers in the future.

— My comments were dismissed or ignored by the Fish and Wildlife Service and the U.S. Geological Survey.

Chapter 5 Forests, Spotted Owls, and the ESA

Bullet Summary
— Conflict over timber harvest and potential impacts to wild-life in the western United States and Alaska was intense in the 1990's and continues.
— Environmentalists claimed that logging and timber harvest negatively impact wildlife.
— Others argued that forests can be managed for multiple-use to produce timber and protect wildlife.
— The focus of the conflict was in the Pacific Northwest where the Endangered Species Act (ESA) listing of the northern spotted owl resulted in major decreases of timber harvest.
— This decrease in timber harvest devastated the timber industry and communities reliant on it.
— The northern spotted owl is a subspecies of the spotted owl species.
— The California spotted owl and Mexican spotted owl are also subspecies of the spotted owl species.
— The ranges of the northern spotted owl and California spotted owl overlap.
— There is some movement and interbreeding of owls among the three spotted owl subspecies
— Subspecies in general, including the northern spotted owl, are not scientifically definitive.
— The northern spotted owl designation as an endangered subspecies is an example of the selective use of equivocal science with the ESA.

Bullet Summary Chapters 1-5: The Endangered Species Act Problems and Solutions

The Endangered Species Act (ESA) has problems with science and policy.

1. Science problems with the ESA are:
   a. The ESA species definition includes subspecies, distinct population segments-DPS, and evolutionarily significant units-ESU which are not scientifically definitive categories.
   b. The risk of extinction expressed as being threatened or endangered, is not scientifically definitive.
   c. What constitutes a significant portion of a species range and the foreseeable future are not scientifically definitive.
   d. Federal government agencies make all of the science determinations for the ESA.

The solution is to:

   a. Change the ESA to apply only to full biological species.
   b. Separate the regulatory and science functions of the ESA so they are not in one agency.
   c. The Fish and Wildlife Service and National Marine Fisheries Service can implement the ESA using science determinations made by others.
   d. The science relevant to the ESA can be done outside government under contract to universities and private sector science companies and institutions, and include rigorous peer review.

2. Policy problems with the ESA are:
   a. The ESA applies on private property.
   b. The federal government takes fish and wildlife management authority from the States and gets deference in court.
   c. The ESA causes a burden of extensive litigation on governments, industries and agriculture, and private landowners.

The solution is to:

a. Change the ESA so it does not apply on private property. Provide voluntary incentives for landowners to protect and enhance habitat for wildlife.

b. Change the ESA to make explicit that States have exclusive jurisdiction over fish and wildlife populations, including subspecies, DPS, and ESU.

c. Change the ESA so the States and federal government have equal co-authority on ESA decisions. Provide incentives for States to protect and enhance habitat for wildlife.

d. End the granting of deference to federal agencies in court, and change or repeal the Equal Access to Justice Act.

Chapter 6 The North Slope of Alaska: Oil Fields and Caribou

Bullet Summary

— Oil and gas fields have been developed on the North Slope of Alaska since the discovery of a major oil reservoir at Prudhoe Bay in 1968 and completion of the Trans-Alaska Pipeline in 1977.

— Concerns over environmental impacts, particularly to fish and wildlife, have been paramount on the North Slope resulting in extensive research and monitoring.

— Impacts of the oil fields on caribou have been a primary concern.

— The Central Arctic caribou herd uses habitats in and around the North Slope oil fields for calving and post-calving summer range.

— Disturbing and displacing caribou from calving areas because of oil field activity has been postulated as a negative impact.

— Some studies show displacement of calving caribou from oil field infrastructure and other studies do not.

— The extent of this impact has been debated, particularly when the studies at Prudhoe Bay are used to predict impacts of oil development in the Arctic National Wildlife Refuge (ANWR).

— The Central Arctic caribou herd grew from 5,000 animals in the late 1970's when the oil fields were first developed to 68,000 in 2010, and declined to 28,000 in 2017.

— Studies of population genetics and population dynamics show that inter-herd movements substantially affect the numbers of caribou in each herd, and data do not support the hypothesis that oil field impacts have caused a population decline.

— The impact of oil fields on the number of caribou in the Central Arctic herd is probably small compared to other factors including winter severity and habitat, immigration into and emigration out of the herd, and calf recruitment.

— Oil field impacts continue to be debated and are used to oppose development in ANWR

— Government reports do not adequately consider published literature on caribou that was sponsored by the oil industry.

Chapter 7 Biology and War

Bullet Summary

— War has been a constant worldwide characteristic of mankind throughout history.

— War is defined as open and declared armed hostile conflict between states or nations.

— War is considered by military scholars as the continuation of policy by other means.

— War is a part of human biology and there are biological causes of war.

— Men have been the combatants in war throughout history.

— The U.S. government instituted full integration of women into all Military Occupational Specialties (MOS) including combat MOS in 2015.

— This government policy was justified because of a perceived need to take full advantage of every individual who can meet military standards.

— Studies showing poorer performance of male-female integrated units than all-male units were dismissed by the government without legitimate justification.

— There is a vast scientific literature on the differences between men and women and why such differences exist.

— Science indicates that the historical pattern of men fighting men and the exclusion of women as combatants in war is likely rooted in basic differences of the sexes that developed as adaptations that enhanced fitness in nature.

— This science was apparently not considered or was dismissed by the government in the decision to integrate women into combat MOS.

— Scientific knowledge can contribute to an understanding of the historical pattern of men fighting men in war, and inform future policy decisions on the roles of the sexes in the U.S. military.

— Changing a historical pattern by allowing women to be integrated with men and fight in war entails risks to military effectiveness and the basic structure and fitness of human populations.
— The risks associated with full integration of the sexes in the military have not been adequately assessed and should be reconsidered in light of biology.

Chapter 8 God and Science

Bullet Summary

— Mankind struggles with the question of whether God exists or does not exist.

— The existence of God is considered a scientific question or a non-scientific spiritual question by different people.

— Some people think God does not exist based on lack of scientific evidence.

— Some people think God does exist based on science.

— We do not know with scientific certainty if God exists or does not exist.

— Science provides strong evidence that humans evolved from ancestors with less mental capacity than modern humans.

— Humans at present may not have the mental capacity to comprehend all aspects of nature, the universe, and God.

— Human evolution included retention of juvenile characteristics (known as paedomorphosis and neoteny) that resulted in larger brains and greater mental capacity in modern humans than in our evolutionary ancestors.

— Future evolution could result in increased mental capacity that allows better comprehension of the universe and God.

— Our current understanding of science and human evolution does not support literal interpretation of creation happening in seven days as described in Genesis.

— Evolutionary analogies to Genesis provide insights regarding creation and the future evolution of mankind's mental capacity to comprehend God.

— Christ telling us to be like children and identifying Himself as the Son of Man is consistent with the evolution of increased mental capacity through retention of juvenile characteristics that can ultimately result in human capacity to understand God.

Chapter 9 Socialist, Communist, Marxist Science, and Lysenkoism

Bullet Summary

— The Soviet Union controlled science to fit the socialist/communist government's agenda during the infamous period of Lysenkoism.
— In the case of Lysenko it was agricultural science.
— Lysenkoism endorsed inheritance of acquired characteristics and rejected Mendelian genetics to be consistent with socialism.
— Lysenko's incorrect science resulted in large failures of Soviet agriculture which resulted in the deaths of millions of people.
— Dissent from Lysenko's theories was outlawed in the Soviet Union in 1948, and dissenters were dismissed from jobs, imprisoned, or sentenced to death as enemies of the state.
— This included an accomplished botanist, Nikolai Vavilov, who died in prison.
— Lysenkoism has been described as a warning of the dangers of bureaucratic and ideological distortions of science.
— Lysenkoism was made possible by a totalitarian socialist/communist government.
— Some Americans are now endorsing socialism in the U.S.
— Modern science in the U.S. is not descending into Lysenkoism, modern science is rigorous, but science can be distorted by political agendas.
— Authors have warned about the threat of Lysenkoism reemerging today for environmental issues such as global warming.
— Americans should be aware of history, and oppose government control of science and stifling dissent to achieve policy agendas.

PART I.
CONFLICTS IN
WILDLIFE SCIENCE

Chapter 1 Introduction: Science, policy, and management

Bullet Summary

— The Endangered Species Act (ESA) is the most powerful environmental law in America.

— The federal government makes scientific determinations for the ESA.

— The only recourse for challenging ESA determinations is to sue in court.

— This is largely ineffective because the government agencies apparently have unlimited funds for legal challenges and receive deference in court.

— This results in a government monopoly of the science and information with the ESA.

— In my experience, ideas, information, and science that are not in agreement with the federal government bureaucracy are dismissed or ignored.

— Prioritizing wildlife conservation and selective use of science has resulted in a bias against natural resource use and agriculture in the government bureaucracy and its associates in academia and environmental groups.

— This is important because environmentalism has become a tool of socialists/communists trying to undermine the basic structure of the U.S. Constitutional Republic.

— Environmental regulation, especially the taking of private property and federal seizure of States' jurisdiction of fish and wildlife with the ESA violates the U.S. Constitution.

— Elected officials, military personnel including veterans, and law enforcement officers, have all sworn an oath to uphold and defend the Constitution.

— Elected officials allowing the ESA to violate the Constitution appear to be dishonoring this oath.

— Several books have documented government misuse of science and policy with the ESA and other regulations (Coffman 1994, Chase 1995, Pombo and Farah 1996, Fitzsimmons 1999, Stirling 2008, Andre 2011, Lyon and Graves 2014, Crockford 2019).

— In this book I provide scientific information and insights that will complement these books.

Chapter 1

INTRODUCTION: SCIENCE, POLICY, AND MANAGEMENT

In his *Letter to the Soviet Leaders*, Aleksandr I. Solzhenitsyn (1974:1) expressed that he had little hope that they, the leaders of the Soviet Union, would consider his ideas. He pointedly noted that he was not in the government and could get no demotion or promotion from them. Therefore his opinions were sincere and without a motive to advance himself, unlike those in the Soviet bureaucracy.

Solzhenitsyn knew the Soviet bureaucracy would not listen to him or acknowledge his ideas. So it is with parts of the U.S. government bureaucracy. My experience is mostly with the environmental bureaucracy, and more specifically, the government fish and wildlife bureaucracy. A difference between the U.S. government and the Soviets is that the U.S. government *does* formally solicit comments and input on proposed rules for endangered species listings, environmental impact statements, and other regulations. Public comments can be provided in written comments, public hearings, and Congressional hearings. This is the official public process integral to our government of, by, and for the people.

However, ideas, information, and science not in agreement with the government bureaucracy are, in my experience, often dismissed or ignored resulting in what I believe is a government monopoly of the science and information used with the Endangered Species Act (ESA) and other environmental issues.

For example, with the ESA the U.S. Fish and Wildlife Service (FWS) and National Marine Fisheries Service (NMFS, also called the National Oceanic and Atmospheric Administration-NOAA Fisheries) write proposed rules and species status reviews, choose peer reviewers, accept or dismiss peer review comments, accept or dismiss public comments, and write final rules to designate endangered species.

The only recourse for challenging such designations is to sue in court. However, FWS and NMFS apparently have unlimited funds for legal challenges and receive "deference" in court as the experts with the correct science simply because they are the federal government. Federal courts defer to an agency's interpretation of a statute or regulation, called Chevron deference or Auer deference, referring to court cases (Supreme Court of the United States 1984, 1997). This has been considered delegation of legislative duties to the executive branch and violates the Constitutional separation of powers. Justice A. Scalia described this practice as tyranny in a water pollution case (Scalia 2013).

Solzhenitsyn's sentiment in his *Letter to the Soviet Leaders* describes how people in the natural resource industries and agriculture, and some scientists like me who work with them, feel when our comments and science are dismissed or ignored by the government bureaucracy. This includes proposed rules for ESA listings, environmental impact statements (often required by the National Environmental Policy Act-NEPA), and wetlands regulation under the Clean Water Act (CWA). I believe this reflects a bias against natural resource development and agriculture (e.g., timber, mining, oil and gas, livestock grazing) in the government bureaucracy, its associates in academia and environmental groups, and the general environmental conservation community.

This is not to say the government bureaucracy is always biased against resource use and agriculture. Many people in the government bureaucracy are honest and unbiased. Sometimes the government bureaucracy will oppose ESA listings or endorse de-listings (i.e., removing a species from the ESA list) when the science indicates they are not endangered with extinction. For example, the Fish and Wildlife Service has supported de-listing wolves and grizzly bears (Federal Register 2017, 2019).

Problems with the ESA are largely because of its expansion from the original intent of preventing extinction of species. The most important expansions of the ESA are inclusion of subspecies and distinct population segments (DPS) in the species definition, protection of ecosystems, and its application on private property. This expansion was done by people commissioned to write the ESA and evidently neither Congress nor the Executive Branch ensured that the ESA as written was consistent with their intent prior to passing it (Mann and Plummer 1995, Lueck 2000). This is a disturbing circumstance where unelected advocates wrote a law with serious consequences for Americans.

Solzhenitsyn's *Letter*, addressing the Soviet socialists/communists, is also relevant because environmentalism has become a tool of socialists/communists trying to undermine the basic structure of the U.S. Constitutional Republic (see Arnold 2007, Horner 2007, Andre 2011). The control of private property with the ESA and CWA is alarmingly consistent with the first plank of the Communist Manifesto: "Abolition of private property in land and application of all rents of land to public purpose." See Chapter 9 regarding socialism/communism and science.

The U.S. Constitution institutes limited federal government power, strong State governments, and individual citizens' liberty and property rights. The 5th Amendment guarantees that private property will not be taken by the government without just compensation, and the 10th

Amendment limits the powers of the federal government while granting most powers to the States and the people.

These principles are violated by environmental regulations that use federal power to control property and behavior. The government regulation and taking of private property without compensation with the ESA is perhaps the most egregious example of this. See Lueck (2000) for a detailed description of the history, politics, and economics of property rights and the ESA. Regulation under the ESA, including control of private property, is often justified with the Interstate Commerce Clause in the Constitution which has been debated extensively (see van Loh 2004, Levin 2013).

The most offensive aspect of the ESA, to me and many other Americans is that it allows the government to regulate private property. After all, the 5th Amendment to the Constitution protects us from the government taking private property. In my discussion of sage grouse in Chapter 2, I suggest that a voluntary incentives program is consistent with American principles and would be more effective in conserving wildlife than government control of private property with the ESA. Thomas (2005) made a similar suggestion, and incentives are widely recognized in conservation (Benjamin 2004).

At the time of this writing, some politicians are openly calling for socialism in the U.S. (see Kesler 2018). However, the absolute federal government power of socialism is antithetical to the U.S. Constitution. Elected officials, military personnel including veterans, and law enforcement officers, have all sworn an oath to uphold and defend the Constitution from foreign and domestic enemies. Our elected officials allowing the ESA to violate the Constitution appear to be dishonoring this oath. Motivating others to honor this oath is one of my goals in this book.

Science, Management, and Policy

It is important to recognize that modern science, including the research done in universities, government agencies, and private industry, is of the highest quality. Our system of peer review, open communication and collaboration, and principled adherence to professional ethics of honesty and rigor assures the best quality basic scientific research. This is perhaps most apparent in the fields of genetics, molecular biology, biochemistry, and biotechnology in which remarkably rigorous science is advancing rapidly (e.g., Mukherjee 2016, Doudna and Sternberg 2017).

The application of science in medicine, agriculture, industry, and engineering is also excellent, and driven by performance and results. The competitive nature of research for funding and recognition, and competition and financial incentives in the private sector for science applications that work and solve real-world problems, assures good quality. For example, consider that the oil and gas industry extracts, refines, and transports millions of barrels and cubic feet of flammable, explosive fluids safely and efficiently every day. There are occasional accidents, but overall the oil and gas industry works incredibly well using science-based engineering. And our agricultural system is astounding considering the science-based production and transport of safe and superabundant food across the nation and the world. See Chapter 9 for a different system in the former Soviet Union where agricultural science was controlled by the government without competition or dissent.

The actual science done by government fish and wildlife biologists is also good. It is innovative and current, using modern technologies such as DNA analyses, satellite Geographic Positioning System (GPS) telemetry, and unmanned aircraft (drones), with rigorous statistical and publication standards. Also, government biologists and resource managers are not always biased against resource use and agriculture. Most people in government research, resource management, and agriculture try to deal with intricate laws and regulations in a helpful and unbiased fashion.

However, in my experience, there is a problem with the selective use of science to advance agendas. It is apparent to me that the objective is often to prevent resource development (e.g., timber harvest, mineral and oil extraction, livestock grazing) using the regulatory power of laws such as the ESA, CWA, and NEPA.

I recognize a potential paradox: I am claiming that American science is high quality, but also contend there is bias in the science used by the government bureaucracy in environmental regulation. A goal of this book is to contribute to raising the standard of the latter to that of the former.

An important insight is that science, policy, and management are not the same. Science does not dictate policy, science informs policy. Science is knowledge concerned with the physical world and its phenomena. Policy is a course of action or a plan. Management is judicious use of means to accomplish an end (Webster's Ninth New Collegiate Dictionary 1988).

A management objective, a standard tool in resource management, would be the end, or goal that a management action is designed to achieve. For example, a management objective could be the desired number of animals in a wildlife population, or the desired yield of corn from a farm. Every environmental issue can be defined by management objectives. Management objectives will differ among people and organizations, and should be clearly stated in every case. For example, some people have a management objective of untouched wilderness for the entire Arctic National Wildlife Refuge (ANWR) in Alaska, and others have a management objective of oil and gas extraction on the northern coastal plain of ANWR. Neither is right or wrong, they are simply different management objectives. What I hope to convey in this book is the importance of honesty in presenting science regardless of one's management objectives.

Considering that science is knowledge, policy is a plan, and management is action to achieve an objective, it is clear that science is not management or policy. For example with the ESA, science can be used to assess if a species is endangered with extinction, policy is a plan to prevent its extinction, and management is the action implemented to prevent its extinction. Keep in mind that science doesn't show anything about whether a human action is right or wrong, it is value-neutral. One can use science to predict the outcome of an action, but science doesn't say what the correct action is.

The ESA includes policies, or a course of action, to prevent extinction of species. To decide if these policies apply in specific situations, science is used to determine if a species is endangered with extinction. The ESA requires use of the "best available science" (see Sullivan et al. 2006) in making these decisions. It seems to me that "best available" is an unnecessary qualifier here. It is self-evident that only available science can be used. Problems arise in deciding what the best science is and who decides what the best science is. With the ESA the best science is decided by the government agencies.

The reader should consult Anderson (2000) for the history, politics, and economics of environmental issues, Lomborg (2001) and Horner (2007) for general environmental issues, and Sanera and Shaw (1996) for explaining environmental issues to children. Other books have done an outstanding job documenting and explaining government misuse of science and policy with the ESA and other regulations (Chase 1987, 1995, Coffman 1994, Fitzsimmons 1999, Pombo and Farah 1996, Stirling 2008, Andre 2011, Lyon and Graves 2014, Crockford 2019). I urge you to read these books. You will be surprised at the amount of well-documented abuse of power by the government bureaucracy and environmental groups. I hope my perspective as a scientist complements these important books.

Chapter 2. The Endangered Species Act: What does Endangered mean?

Bullet Summary

— The Endangered Species Act (ESA) is the most powerful environmental law in America.

— The ESA definition of wildlife species includes species, subspecies, and distinct population segments (DPS).

— This expands the ESA beyond its original scope of preventing extinction of species.

— Species may be listed under the ESA as either endangered or threatened with extinction.

— An endangered species is a species which is in danger of extinction throughout all or a significant portion of its range.

— A threatened species is a species which is likely to become an endangered species (i.e., likely to become endangered with extinction) within the foreseeable future throughout all or a significant portion of its range.

— It is important to recognize that the ESA applies to species that are at risk of extinction.

— Extinction means there are none (zero) remaining.

— The ESA should not apply to species that are simply experiencing declines in number or habitat and not at risk of extinction.

— Determinations of whether species are endangered or threatened with extinction are not scientifically definitive and made by government agencies.

— What constitutes a significant portion of a species' range and what constitutes the foreseeable future are not scientifically definitive and are determined by government agencies.

— An example is the greater sage grouse that was found to be warranted for ESA listing in 2010.

— The greater sage grouse was not endangered with extinction but the ESA listing was justified with predictions and speculation regarding its future habitat and numbers.

— The greater sage grouse was reassessed and found not warranted for ESA listing in 2015.

— The ESA allows the government to regulate use of private property. A voluntary incentives program for landowners instead of government regulation with the ESA would be more effective in protecting wildlife and consistent with American property rights.

Chapter 2.

THE ENDANGERED SPECIES ACT: WHAT DOES "ENDANGERED" MEAN?

President Ulysses S. Grant in his Inaugural Address, 4 March 1869, stated:

> "I know of no method to secure the repeal of bad or obnoxious laws so effective as their stringent execution."

President Grant's words reflect wishful thinking shared by many Americans who are negatively impacted by the Endangered Species Act (ESA) which has been called the most powerful environmental law in America. The ESA empowers the federal government to regulate private and public property and has been very controversial with extensive litigation. Several books have documented government misuse of science and policy with the ESA and other regulations. These books should be consulted for understanding the past implementation of the ESA (Coffman 1994, Chase 1995, Pombo and Farah 1996, Fitzsimmons 1999, Stirling 2008, Andre 2011, Lyon and Graves 2014, Crockford 2019).

The ESA has greatly increased the federal government's authority over fish and wildlife, in large measure because of the expansion of the scope of the ESA by those commissioned to write it (Mann and Plummer 1995, Lueck 2000). The most important expansions included subspecies and distinct population segments (DPS) in the definition of species, and regulation of private property. The ESA was also expanded to include the nebulous goals of preserving biodiversity and the ecosystems on which endangered species depend. Jack Ward Thomas, former Chief of the U.S. Forest Service, discusses these topics in an interview in *Range Magazine* (Thomas 2005). These expansions of the ESA make it a powerful land management tool of the government, especially because it applies on federal, State, and private land.

Citizens, environmental groups, or a government agency can petition the government to designate an endangered species under the ESA. To determine if the ESA is appropriate, the first step is to determine if the petitioned species is a species as defined in the ESA (see Chapter 3). The second step is to determine if the species is threatened or endangered with extinction. To be designated "endangered" and listed under the ESA, a species must be endangered with extinction throughout all or a significant portion of its range. To be designated "threatened" and listed under the ESA, a species must be likely to become endangered with extinction in the foreseeable future (often considered to be 30 to 100 years) throughout all or a significant portion of its range. A species' range refers to the area over which it occurs.

The meaning of the word extinct is clear as defined in Webster's Ninth New Collegiate Dictionary (1988): "no longer existing". The terms "in danger of extinction", "likely to become endangered", the "foreseeable future", and a "significant portion of its range" are not empirically defined and their designation is left to the judgement of the government agencies.

In my opinion these designations are often speculative. An example is polar bears (see Chapter 4) in which computer modeling was used to predict that the entire polar bear species was likely to become endangered with extinction in the foreseeable future and was listed as a threatened species under the ESA. Other species considered for ESA listing with which I am familiar are discussed below. The information below on sage grouse, bison, grizzly bears, and wolves is summarized from articles previously published in the *Western Livestock Journal* (Cronin 2019a, 2019b, 2019c, 2019d).

Prairie Dog

An example of an unwarranted ESA listing is the black-tailed prairie dog. In 2000, the Fish and Wildlife Service determined that listing the black-tailed prairie dog as a threatened species was warranted (Federal Register 2000). The listing was precluded by other higher priority listings, but the species was considered threatened with extinction and put on the "candidate list". Subsequent analysis and input from States and others resulted in estimation of a population of 18,420,000 black-tailed prairie dogs in the United States. The species was clearly not near extinction. Considering this information, the Fish and Wildlife Service changed the decision and determined that the petition to list the black-tailed prairie dog was not warranted and removed it from the candidate list in 2004 (Federal Register 2004).

Around the same time, scientists with the U.S. Department of Agriculture (USDA), Montana State University, and South Dakota State University published a seminal paper (Vermeire et al. 2004) in which they showed that many of the claims used to support the proposed ESA listing of the black-tailed prairie dog were not valid. Vermeire and co-authors called for objective evaluation of all relevant science. They recognized that selective interpretation of data and literature based on personal values resulted in different opinions of the biology and status of prairie dogs. Vermeire et al.'s presentation of science, field observations, and knowledge in the agriculture community clarified the issues about prairie dogs and probably contributed to the decision not to list them under the ESA.

Sage grouse

The Fish and Wildlife Service determined the greater sage-grouse was warranted for ESA listing as an endangered or threatened species in 2010, but delayed listing it because of higher priorities. This threat of an ESA listing resulted in extensive work by ranchers, counties, States, and others, which in turn resulted in a not warranted listing decision in 2015 (Federal Register 2010, 2015). This work probably helped greater sage-grouse populations. But was it justified by good science, or was the 2010 warranted for listing decision unjustified?

I published a paper in *Rangelands*, a journal of the Society for Range Management (Cronin 2015), in which I describe problems with the science used in the 2010 greater sage-grouse ESA decision. One consideration is that the entire greater sage-grouse species was considered threatened or endangered with extinction. However, Gunnison sage-grouse are listed as a threatened species separately from the greater sage-grouse (Federal Register 2014) but whether these are actually different species is questionable. Also, a DPS of the greater sage-grouse in the Mono Basin of California was found not warranted for ESA listing in 2015, separately from the ESA decision for the entire greater sage-grouse species. Such splitting of species, subspecies, and DPS greatly complicates the ESA (see Chapter 3).

The essential question for the 2010 ESA listing decision was if the entire greater sage-grouse species was actually threatened or endangered with extinction. The Fish and Wildlife Service found this to be the case in their 2010 determination. This is an extreme prediction, considering that in 2010, greater sage-grouse occupied 56 percent of their historic range in 11 states and two Canadian provinces and there was an estimated population of 535,542 birds. Although there may be problems with impacts to habitat, predation, and other factors as with all wildlife, sage-grouse were not endangered with extinction.

The 2010 endangered finding was based on predictions of future impacts and habitat loss from population models that have inherent uncertainty and questionable assumptions. One such prediction was reduced fitness due to inbreeding in populations isolated by development that would contribute to the risk of the species' extinction. This is quite speculative. In the models many sage-grouse populations were actually found *not* likely to go extinct, so logic indicates the entire species was not likely to go extinct.

Despite this, the Fish and Wildlife Service found that sage-grouse were threatened or endangered with extinction anyway. To reach this finding the Fish and Wildlife Service used published model results that did not predict extinction but then "anticipated" future impacts and habitat conditions and concluded the species was threatened or endangered with extinction.

Instead of speculating about future impacts, the Fish and Wildlife Service should have posed their predictions as hypotheses to be tested with observations, and not make unfounded conclusions. The greater sage-grouse might have needed better management and habitat protection in some areas, but I believe it was simply speculation that concluded it was threatened or endangered with extinction. In my opinion, using inconclusive science and speculation to support an ESA listing is not the best available science.

It is important to recognize that proactive management can be better than invoking the ESA for species such as sage-grouse. For example, controlling predators and rearing sage-grouse in captivity for release into wild populations can benefit local populations (Oesterle et al. 2005, Orning 2014). Voluntary incentives through programs like the conservation reserve program (CRP) have also succeeded in improving wildlife habitat and populations (Eggebo et al. 2003). Such voluntary incentive-based programs are more likely to be successful at conserving sage-grouse than mandatory ESA regulation, without infringing upon multiple-use principles and property rights. I believe voluntary incentives instead of federal regulation with the ESA on private property would be a major improvement for both the nation and wildlife.

Bison

Bison (also called buffalo) have been under consideration as a threatened or endangered species with the ESA. Two subspecies of bison have been recognized. The wood bison subspecies historically occurred in northwest Canada and Alaska and the plains bison subspecies occurred throughout much of North America. However, there is disagreement on the validity of the bison subspecies designations (Geist 1991, Cronin et al. 2013a), and the subspecies category is not scientifically rigorous (see Chapter 3).

The subspecies designations are important because the wood bison subspecies is listed as a threatened species under the ESA and a population of the plains bison subspecies was petitioned for ESA listing in 2018, after previous petitions in 2007, 2014, and 2015. These petitions were initially denied by the Fish and Wildlife Service but they are reconsidering the 2018 petition following a judge's order (Cooper 2018). The 2018 ESA petition is to designate the plains bison in Yellowstone National Park as a distinct population segment (DPS).

Wood bison are thought to have historically occurred in Alaska, but went extinct there a few hundred years ago (Gates et al. 2010). Wood bison from herds in Canada were moved to Alaska in the early 2000's. The animals brought to Alaska from Canada included 13 that were owned by a rancher, Scott Hollembaek, in interior Alaska. These bison were confiscated by the federal government in 2003 because wood bison are listed under the ESA. Evidently, importation and private ownership of an ESA-listed species, or in this case a subspecies, is illegal. Hollembaek lost his investment of $25,000 to buy these animals, his time, effort, and transportation costs, and was also fined $18,000 by the government. Scott and his wife Ruby operate a top-notch ranch with bison and elk (the Alaska Interior Game Ranch) that enhances wildlife conservation, yet he was penalized because of the ESA. In this case the ESA had a negative effect on wildlife, and violated American citizens' property rights.

Hollembaek's bison were transferred to State of Alaska ownership and combined with 52 others brought to Alaska from Canada in 2008. The State of Alaska imported the 52 wood bison from Canada, so it's apparently OK for government to import and possess wood bison, but not American citizens. These animals were kept in captivity at the Alaska Wildlife Conservation Center where the herd grew until 2015, when 100 of these wood bison were released into western Alaska.

Bison were close to extinction in the late 1800's, but were conserved through the efforts of ranchers, sportsmen, Native Americans, and the U.S. and Canadian governments. Bison have been restored in several public and private herds and there are now about half a million bison in North America (Gates et al. 2010). Recent estimates include 20,504 plains bison in public herds in 2008, about 400,000 bison on 6,400 farms and ranches in U.S. and Canada in 2010, and about 11,000 wood bison in Canada in 2013. As described above, wood bison have also been transplanted to Alaska.

These numbers indicate that bison are not endangered with extinction. The 2018 ESA petition for the Yellowstone plains bison herd is based on designating it as a DPS, separate from all of the other plains bison. A primary justification of the ESA petition is the claim that the Yellowstone bison are free of cattle genes. Interbreeding of bison and cattle in the past resulted in the transfer of genes from cattle into bison, and some bison still have residual genes from this interbreeding. The importance of this is not scientifically established because the assessment of cattle genes requires extensive genetic analysis, and some herds have not been tested. For example, there are plains bison in Alaska that are descendants of bison introduced from Montana in the 1920's that may also be free of cattle genes.

Regardless, herds other than the Yellowstone bison are also free of, or have small amounts of, cattle genes (Halbert and Derr 2007, Hedrick 2009, Ranglack et al. 2015) and the effect of small amounts of cattle genes on an animal's bison characteristics is uncertain. The claim that bison with a small percentage of cattle genes don't qualify as bison, particularly for the ESA, is not scientifically established.

There are other considerations for the Yellowstone bison as an endangered DPS. The Yellowstone herd is infected with brucellosis which is a threat to livestock and other wildlife. There are many other bison herds that are not infected and can be used for conservation. Also, the Yellowstone herd was estimated at 4,527 bison in 2018 and is not endangered, or likely to become endangered, with extinction. The National Park Service planned to capture hundreds of bison in Yellowstone in 2019 for slaughter because there are too many bison for the available habitat. The park tries to maintain a herd of 4,200-4,500 head.

Grizzly Bear

Grizzly bears are common in Alaska and parts of Canada. They are also called brown bears depending where they live. The large bears of coastal areas and islands of Alaska including Kodiak Island are called brown bears, while inland populations are called grizzly bears, but they are the same species. Brown bears also occur in Europe and Asia. Polar bears in the Arctic, and black bears which range across the U.S., are the other bear species in North America.

Grizzly bears and polar bears are closely related species and may occasionally interbreed where their ranges overlap in the Arctic. However, this is rare and they are different species (Cronin et al. 1991, Cronin and MacNeil 2012, Cronin, et al. 2013b, Cronin, et al. 2014). Other closely-related species occasionally interbreed, including mule deer and white-tailed deer; and wolves and coyotes.

The status of grizzly bears and the ESA in the western States is complicated, and as of 2019 is:

Alaska: Not listed as an endangered species in Alaska. There are about 30,000 grizzly/brown bears in Alaska.

Mexico: Listed as an endangered subspecies (the Mexican grizzly bear), although the Mexican grizzly might be extinct.

Yellowstone National Park: Listed as a threatened distinct population segment (DPS) in and around Yellowstone National Park (called the Greater Yellowstone Ecosystem). This population was designated a DPS in 2007, delisted in 2017, and relisted by a judge in 2018. There are about 700 bears in this population.

The Contiguous 48 States (i.e., all of the United States except Alaska and Hawaii): Listed as a threatened DPS in the contiguous 48 States in 1975. The Yellowstone DPS was split from the Contiguous 48 States DPS in 2007. The Contiguous 48 States DPS currently consists of five Recovery Zones: The northern Rocky Mountains in Montana, called the Northern Continental Divide Ecosystem, with about 765 bears; The Selkirk Mountains in northeast Washington, northern Idaho, and adjacent British Columbia, Canada with about 70-80 bears; the Northern Cascade Mountains in north-central Washington with less than 20 bears; the Bitterroot Mountains in southwest Montana and eastern Idaho where it is not known if there are currently any grizzly bears there but one was seen in 2002 and tracks were seen in 2012, and it is considered as potential habitat and an experimental population;

and the Cabinet-Yaak area of northwest Montana and northern Idaho where there are about 45 bears.

The ESA listings for grizzlies thus include one Contiguous 48 States DPS with five recovery zones, the Yellowstone DPS, and the Mexican grizzly subspecies. The scientific uncertainty of the subspecies and DPS categories is described in Chapter 3.

Wolf

Wolves are listed under the ESA as various species, subspecies, and DPS. An excellent review of wolves and the ESA is provided by Lyon and Graves (2014). The current status of wolves and the ESA is:

Gray wolf: Not listed in Alaska.

Gray wolf: The Contiguous 48 states distinct population segment (DPS). Listed as endangered since 1978. This includes 44 states. Of the other six of our 50 United States, four (Minnesota, Wyoming, Idaho, Montana) have their own ESA-listed wolves, wolves never occurred in Hawaii and wolves are not listed under the ESA in Alaska.

Gray wolf: Northern Rocky Mountain distinct population segment (DPS). Delisted in 2011 by Congress, with various re-listings and final delisting by 2017. Includes Montana, Wyoming, Idaho, and parts of Washington, Oregon, and Utah. It was originally part of the contiguous 48 states DPS and split into its own DPS in 2011. Ted Lyon describes the efforts to get Congress to delist this DPS in 2011 in the book *The Real Wolf* (Lyon and Graves 2014).

Gray wolf: Western Great Lakes distinct population segment (DPS). Delisted in 2011 and relisted as threatened in 2015. Includes Minnesota and parts of Wisconsin and Michigan. It was originally part of the contiguous 48 states DPS and split into its own DPS in 2011. Some wolves in the Great Lakes DPS contain coyote DNA from interbreeding (i.e., hybridization) of wolves and coyotes (see Cronin and Mech 2009).

Mexican gray wolf: Subspecies. Listed as endangered in 2015. Includes the southern half of Arizona and New Mexico. Its historic range has been suggested to extend into Texas, California, Colorado and Utah. It was originally part of the contiguous 48 states DPS and listed separately as a subspecies in 2015.

Alexander Archipelago gray wolf: Subspecies. Found not warranted for listing as a subspecies in 2016. Occurs in southeast Alaska and adjacent British Columbia, Canada.

Red wolf: Species. Listed as an endangered species in 1967. The original range was the southeast U.S. It is now restricted to an experimental population in North Carolina.

The eastern wolf in the eastern States and Canadian Provinces has also been identified as a species (Chambers et al. 2012) but is not listed under the ESA. The only existing population in Ontario, Canada has been found to have considerable amounts of coyote DNA from hybridization of wolves and coyotes.

Note that the gray wolf is not listed as a species. It is split into subspecies and DPS. If the original contiguous 48 states DPS is considered without splitting off the Great Lakes DPS and Northern Rockies DPS, it is not endangered, because both DPS have been delisted (although the Great Lakes DPS was relisted). Logic indicates that if these DPS aren't endangered, the entire 48 state DPS can't be endangered. But the Fish and Wildlife Service split these DPS for separate consideration, leaving the contiguous 48 states DPS still listed.

There has been debate about whether the red wolf (and the eastern wolf) should be designated as a species or as a subspecies. The National Academies of Sciences, Engineering, and Medicine (2019) and Chambers et al. (2012) support species status for the red wolf. There is also debate about whether the Alexander Archipelago wolf and the Mexican wolf should be designated subspecies. These subspecies designations are supported by some authors (Fredrickson et al. 2015, Weckworth et al. 2015, National Academies of Sciences, Engineering, and Medicine 2019) and not supported by others (Cronin et al. 2015a, 2015b).

There is inconsistency in the use of wolf subspecies with regard to the ESA. The Northern Rockies wolf DPS range in Montana, Wyoming, and Idaho was originally occupied by the plains wolf subspecies, but the wolves introduced into this area in the 1990's were from Canada in the range of the northern wolf subspecies. That is, a non-native subspecies was put into the Northern Rockies. The enthusiasm about subspecies with which ESA advocates endorse the Mexican wolf and Alexander Archipelago wolf was not seen during the introduction of the "wrong" subspecies into the Northern Rockies.

Wolves, grizzly bears, and bison are examples of ESA listings that resulted from splitting a species into subspecies and DPS. The science on such classifications is explained in Chapter 3.

Chapter 3. The Endangered Species Act: What is a Species?
Bullet Summary

— The Endangered Species Act (ESA) definition of "species" includes species, subspecies, and distinct population segments.

— This greatly expands the scope of the ESA beyond preventing extinction of species.

— There are several scientific definitions of species, but species are generally recognized as interbreeding populations that cannot successfully interbreed with other species.

— Wildlife species are generally identifiable, although some species designations are uncertain.

— Subspecies and distinct population segments (DPS) are populations within a species, that is, a part of a species, not the entire species.

— Subspecies are populations in a geographic area that differ from populations in other areas in traits such as size, color, and genetics.

— Subspecies are not scientifically definitive and subspecies designations are often subjective.

— Subspecies determinations for the ESA are made by government agencies.

— Distinct population segments (DPS) are populations designated considering their discreteness and significance.

— Discreteness refers to a DPS's differences from other populations. It is similar to the criteria that distinguish subspecies. DPS can also be separated by international borders.

— Significance refers to the importance of a DPS to the entire species and if it is unique in various ways.

— Discreteness and significance are not scientifically definitive.

— Discreteness and significance determinations for the ESA are made by government agencies.

— DPS are not scientifically definitive.

— DPS determinations for the ESA are made by government agencies.

— DPS for Pacific salmon have been named evolutionarily significant units (ESU).

— ESU determinations are not scientifically definitive.

— ESU determinations for the ESA are made by government agencies.

Chapter 3.

THE ENDANGERED SPECIES ACT: WHAT IS A SPECIES?

The Endangered Species Act (ESA) has been called the most powerful environmental law in America. A primary reason the ESA is so powerful is because it allows the federal government to consider fish and wildlife *populations* as *species*. This is because the ESA definition of species includes species, subspecies, and distinct population segments (DPS). Subspecies are populations in geographic areas (i.e., subdivisions of a species) and DPS are also populations in geographic areas (i.e., subdivisions of a species), so a major expansion of the scope of the ESA from its original intent was to include populations in addition to entire species.

Considering the name of the ESA, it is reasonable to assume that the original intent of Congress and the President in passing the ESA was to protect entire species, not subspecies and DPS. Otherwise, Congress would have named it the Endangered Species, Subspecies, and Population Act. However, as of 2006 more than 70% of the mammals listed under the ESA in the U.S. were subspecies or DPS, not entire species (Cronin 2006).

The scope of the ESA was expanded to include subspecies and DPS by those commissioned to write it and apparently neither Congress nor the Executive Branch ensured the ESA as written was consistent with their intent (Mann and Plummer 1995, Lueck 2000). This results in two important considerations. First, designation of subspecies and DPS are not quantitative and are quite subjective. This allows ESA consideration of any local population of fish and wildlife. Second, species, subspecies, and DPS designations for the ESA are made by government agencies, who decide what constitutes the best available science.

This chapter focuses on science relevant to what constitutes a species under the ESA including species, subspecies, distinct population segments (DPS), and evolutionarily significant units (ESU, which are DPS of salmon in Pacific Ocean watersheds). Other terms have been suggested for wildlife populations, but all of these terms simply refer to populations with various qualifiers (Cronin 2006, Cronin 2007a).

The word "population" is used in assessing species, subspecies, and DPS, so understanding its meaning is important. Scientific definitions can be found in textbooks (e.g., Bergstrom and Dugatkin 2016, Futuyma and Kirkpatrick 2017), and a simple definition is:

> A population is a group of animals or plants in a geographic area that is more or less separate from groups of the same species in other areas.

It is also important to understand the science of biological classification known as taxonomy. Taxonomy is based on phylogeny, which is the evolutionary history of a group's ancestry and descent. Species and subspecies are part of the taxonomic system and should be designated based on phylogeny, as described below.

Species

There is a well-known "species problem" in biology: species are difficult to define and identify in a rigorous fashion (Mayr 1957). There are several scientific definitions of species, based on different species concepts (see Bergstrom and Dugatkin 2016, Futuyma and Kirkpatrick 2017). The most common definition is based on the biological species concept (BSC, Mayr 1982) in which species are groups of populations that can interbreed with each other, but cannot interbreed with other such groups.

Other species concepts emphasize evolutionary relationships, such as the phylogenetic species concept (PSC, Cracraft 1989). A phylogenetic species is a group of organisms that is monophyletic and has shared derived characters. The PSC is confusing to many people and not particularly useful because phylogenetic species can be inconsistent with species traditionally designated with the BSC, including most wildlife.

A more useful definition is the evolutionary species concept (ESC, Simpson 1961, Wiley 1978). An evolutionary species is a lineage of populations with its own evolutionary history which is reproductively isolated from other lineages. This is essentially an elaboration of the BSC to include common ancestry. This is consistent with species being a taxonomic category, and taxonomy being based on phylogeny. The point is that plants and animals are classified in taxonomy according to their ancestry. Groups with common ancestry will be classified into groups at various levels. For example, wolves and coyotes have a relatively recent common evolutionary ancestry so in the taxonomic system they are both in the Canid family (Canidae) and the genus *Canis*, but they are different species; wolf *Canis lupus* and coyote *Canis latrans*.

Another useful species concept is the genetic species concept (GSC, Baker and Bradley 2006). A genetic species is a group of interbreeding, genetically compatible natural populations that are genetically isolated from other such groups.

A focus on genetic isolation rather than reproductive isolation distinguishes the GSC from the BSC. The difference between genetic and reproductive isolation is subtle and described by Baker and Bradley (2006).

Genetic isolation of species is consistent in the BSC, ESC, and GSC with the general recognition of species as groups that don't successfully interbreed. The literature on species is vast (see Mayr 1982, Bergstrom and Dugatkin 2016, Futuyma and Kirkpatrick 2017) and should be consulted for a full understanding of species concepts.

Because different species concepts and definitions can result in different species designations, it is necessary to identify what species concept one is using. This is particularly important for species considered under the ESA because of the potential for disagreement over species designations.

There is agreement on most fish and wildlife species designations by State fish and game departments and professional scientific societies (e.g., the American Society of Mammalogists). However, there are some uncertain species designations that are important because of the ESA.

Designation of wolf species is an example. The red wolf has been recognized by the U.S. Fish and Wildlife Service (Chambers et al. 2012) and the National Academies of Sciences, Engineering, and Medicine (2019) as a separate species from the gray wolf. The eastern wolf has also been recognized as a species (Chambers et al. 2012).

In my opinion, these species designations are questionable. The red wolf in the southeastern U.S. and eastern wolf in the northeastern U.S. and Canada differ from gray wolves in some morphological and genetic characters. The morphological differences are primarily in body and skull dimensions and coat color. Genetic data do not show absolute differences between the red wolf, eastern wolf, and gray wolf; and the red wolf and eastern wolf have interbred with coyotes which complicate genetic analyses. In addition, there were originally contiguous wolf populations across North America with no impassable barriers to movements. This makes it hard to conceive of reproductive isolation occurring among wolf populations. I don't think there was reproductive isolation or independent ancestries that warrant species designations for red wolves, eastern wolves, and gray wolves but others do (Chambers et al. 2012, National Academies of Sciences, Engineering, and Medicine 2019). It is important to recognize that in cases where reproductive isolation and independent ancestry are not definitive, judgements are necessary when designating species (Avise 2000). This means that in some cases, such as the wolves, species designations might be scientifically equivocal.

Subspecies

Many high-profile "species" listed under the ESA are actually subspecies (e.g., the Mexican wolf, northern spotted owl, Preble's meadow jumping mouse, wood bison, and coastal California gnatcatcher). There is often disagreement about the legitimacy of the entire subspecies category and individual subspecies designations (e.g., Zink 2004, Cronin et al. 2015a, 2015b, Fredrickson et al. 2015, Weckworth et al. 2015, Patten and Remsen 2017).

Important considerations of subspecies include:

Phylogeny (i.e., evolutionary relationships) is the basis of taxonomy and organisms are classified in accordance with common ancestry.

Subspecies is a taxonomic category, designated with a Latin trinomial name

Subspecies should be classified based on phylogeny.

Population genetics processes, including recent common ancestry, selection, genetic drift, and gene flow, can confound phylogenetic inference of putative subspecies.

Subspecies designations are often partially or entirely based on morphology which can be influenced by the environment and may not reflect phylogeny.

Subspecies is synonymous with the term "race".

The evolutionary biology community has largely dismissed the subspecies (and race) category as subjective and not scientifically rigorous.

Conservation biology efforts and the ESA are often focused on subspecies despite the subjectivity of the subspecies category.

Subspecies are generally considered as populations within a geographic subdivision of the range of a species that are phenotypically similar and differ taxonomically from populations of the species in other geographic areas (Mayr 1982).

Another definition considers subspecies as populations that are phylogenetically distinguishable from, but are able to interbreed with, other populations of the same species. Several, genetically based traits should provide evidence of phylogenetic distinction (Avise and Ball 1990). This definition reflects the explicit recognition of phylogeny as the primary criterion for taxonomy, including subspecies.

However, because subspecies are members of the same species they can interbreed, and strict phylogenetic distinctiveness among populations of the same species simply may not occur. That is, the population genetics processes of recent common ancestry, selection, genetic drift, and gene flow at present or in the recent past between individuals of different subspecies may prevent phylogenetic differentiation. The inherent subspecies characteristic of being members of the same species can prevent subspecies from having the basic criterion used for taxonomic classification: phylogenetic distinctiveness.

Several prominent authors have commented on the lack of rigor of subspecies and race categories. The entire assessments of the authors cited below should be read for a full understanding of the topic. However, they show a consistent view of the subjective nature of subspecies and races.

Wilson and Brown (1953) criticized the lack of rigor in designating subspecies and questioned their validity (Wilson [1994] later changed his view and acknowledged that subspecies can be useful in describing variation within species). Futuyma (1986) noted that there are no specific criteria for the extent of differentiation of populations for designation as subspecies, and that some biologists believe subspecies designations should be abandoned. Haig et al. (2006) observed that there is considerable disagreement about definitions of subspecies, found no consensus on a subspecies definition in a review of the literature, and felt that the scientific community is somewhat comfortable with the subjectivity of subspecies. Avise (2000:308) noted that in some cases subspecies and species are not definite, and judgments are necessary when designating them. Ehrlich (2000:49, 291) described how species can be split into subspecies by selecting different traits with which to classify them, and that the subspecies concept is no longer used in most evolutionary biology literature. Other authors have also described the subjective nature of subspecies and races (Mayr 1954:87, 1970, Vanzolini 1992, Cavalli-Sforza et al. 1994:19, O'Gara 2002, Zink 2004).

Regarding the point made by Ehrlich (2000), we might ask: is the subspecies category no longer used in the evolutionary biology literature? Without reviewing the extensive evolutionary literature, I simply note that subspecies is not mentioned in the text of two commonly-used textbooks on evolution. One of them (Bergstrom and Dugatkin 2016), makes only brief mention of the lack of genetic support for the designation of human races. The other textbook (Futuyma and Kirkpatrick 2017) mentions subspecies and race only in the glossary as populations with distinctive features in a specific geographic area and note that race is poorly defined and synonymous with subspecies.

It appears that the subspecies category is not prominent, and considered as poorly defined in the evolutionary literature, although it is still commonly used in the other biological literature (e.g., wildlife and conservation biology).

Subspecies and the Endangered Species Act

For the purpose of this chapter it is important to reconcile the lack of scientific support for subspecies with the continued use of subspecies, particularly in wildlife and conservation biology.

The most obvious reason for the continued use of subspecies is that some biologists consider subspecies a legitimate category and think that some populations qualify to be designated as such (see National Research Council 1995, Chambers et al. 2012, National Academies of Sciences, Engineering, and Medicine 2019). Despite the negative aspects of the subspecies category, biologists recognize that there is important geographic variation within species, and the subspecies concept can be useful in scientific discourse. In these cases I suggest that biologists acknowledge the inherent subjectivity of the subspecies category and that dissenting views of subspecies designations are legitimate.

A second reason why subspecies are used may be provided by Dawkins (2004:300), who coined the term "tyranny of the discontinuous mind" to describe the tendency of people to designate non-distinct groups with discrete names. Dawkins was writing about species evolving over space and time and their artificial classification into discrete, named groups. Dawkins' recognition of the inherent tendency of people to force things into discrete groups whether warranted or not, is comparable to some subspecies designations.

A third reason for the continued use of subspecies is that named subspecies are more likely to benefit from conservation efforts than populations without a formal name (Rojas 1992, Patten and Campbell

2000). This is apparently the case with many ESA listings of subspecies including the high-profile cases of the northern spotted owl, coastal California gnatcatcher, Preble's meadow jumping mouse, wood bison, and Mexican gray wolf.

The phenomenon of splitting species and subspecies has been recognized as "species inflation" (The Economist 2007, Marris 2007), with the implication that species, subspecies, and other intra-specific groups have been designated to justify conservation efforts. Splitting subspecies is particularly common with the ESA and has resulted in scientists taking definitive stances for or against subspecies designations despite their subjectivity (see Cronin 2007b, Cronin and Mech 2009, Cronin et al. 2015b, Fredrickson et al. 2015, Weckworth et al. 2015).

The ESA requires that the best available science is used in listing determinations. Given the preceding discussion one can ask: is including the subspecies category in the ESA species definition the best available science? In my opinion it is not because of the arbitrary and subjective nature of the entire subspecies category described above. This opinion is consistent with those in the evolutionary biology community who consider subspecies as not scientifically rigorous. Some scientists consider the subspecies category scientifically defensible, but others do not. Likewise, there is seldom a consensus on individual subspecies designations. Unresolvable differing views on the scientific validity of subspecies should exclude subspecies from the ESA species definition.

Avise's (2000) insight that judgements are necessary for subspecies designations, leads to an important question: Who makes the judgements on subspecies designations? Who decides what is and isn't a subspecies considering the best available science? For the ESA it is the federal government agencies that implement the ESA: the Fish and Wildlife Service and National Marine Fisheries Service. This puts these agencies in the difficult position of determining the status of subspecies, a concept that is not supported by many scientists.

It appears to me that although the evolutionary biology community recognizes the subjectivity of the subspecies (and race) category, some in the wildlife biology/conservation community advocate it to enhance conservation efforts without adequately acknowledging its lack of scientific rigor. Although conservation is a laudable goal, selective use of science is not a laudable way to achieve it. This is especially the case when people's personal and economic well-being is affected. The economic and employment concerns of people negatively affected by ESA regulation are no less laudable, and warrant equal or greater consideration, than that of conservation concerns.

A solution is for scientists to continue using the subspecies category if they choose, while acknowledging its shortcomings. However, because of the well-documented subjective nature of subspecies, it should not be used in conservation and management policies, such as the ESA, that have economic and legal ramifications (see Geist 1992). Instead, populations (including putative subspecies) can be identified as management/conservation units without the encumbrances of subspecies (Cronin 2006, 2007b).

Wolf subspecies designations have been controversial and extensively debated regarding ESA listings. One case is the wolf in southeast Alaska which is considered a subspecies by some biologists but not by others (Chambers et al. 2012, Cronin et al. 2015a, 2015b, Weckworth et al. 2015). See Chapter 5 for more information on the wolf subspecies in southeast Alaska. Another case is the Mexican wolf subspecies, which is listed under the ESA. The National Academies of Sciences, Engineering, and Medicine (2019) concluded that the Mexican wolf is a valid subspecies, although the science behind this is not definitive (see Cronin et al. 2015a, 2015b, Fredrickson et al. 2015).

Other cases of subspecies listed under the ESA have also caused controversy. The coastal California gnatcatcher was designated a subspecies with appallingly poor data and subsequently shown to be not warranted as a subspecies with rigorous genetic and statistical analyses (Cronin 1997, Zink et al. 2000, 2014, Skalski et al. 2008). It remains listed under the ESA.

Similarly, the Preble's meadow jumping mouse was listed as an endangered subspecies, but genetic data indicate it does not warrant subspecies designation (Ramey et al. 2005, 2006, 2007, Malaney and Cook 2013). It remains listed under the ESA.

The northern spotted owl is perhaps the best known subspecies listed under the ESA. Its subspecies status is not definitive as described in Chapter 5. It remains listed under the ESA.

Two subspecies of bison (also called buffalo) have been recognized under the ESA. Wood bison historically occurred in northwest Canada and Alaska, and plains bison occurred across much of North America. Following near extinction, bison have been restored to several public and private herds (Gates et al. 2010). However, there is not agreement on the validity of the bison subspecies designations (Geist 1991, Cronin et al. 2013a). This is important because the wood bison subspecies is listed under the ESA and a DPS of the plains bison subspecies was petitioned for ESA listing in 2018. See Chapter 2 for additional information on bison and the ESA.

Distinct Population Segments and Evolutionarily Significant Units
As discussed above, the ESA species definition includes distinct population segments (DPS). Congress directed that distinct population segments be listed "sparingly", but this has not been the case and there are many DPS listed under the ESA (Cronin 2006).

The federal government has a DPS policy (Federal Register 1996):

"Three elements are considered in a decision regarding the status of a possible DPS as endangered or threatened under the Act...

1. Discreteness of the population segment in relation to the remainder of the species to which it belongs;
2. The significance of the population segment to the species to which it belongs, and
3. The population segment's conservation status in relation to the Act's standards for listing (i.e., is the population segment, when treated as if it were a species, endangered or threatened.

Discreteness: A population segment of a vertebrate species may be considered discrete if it satisfies either one of the following conditions:

1. It is markedly separated from other populations of the same taxon as a consequence of physical, physiological, ecological, or behavioral factors. Quantitative measures of genetic or morphological discontinuity may provide evidence of this separation.
2. It is delimited by international governmental boundaries within which differences in control of exploitation, management of habitat, conservation status, or regulatory mechanisms exist that are significant in light of section 4(a)(1)(D) of the Act.

Significance: If a population segment is considered discrete under one or more of the above conditions, its biological and ecological significance will then be considered in light of Congressional guidance (see Senate Report 151, 96th Congress, 1st Session) that the authority to list DPS's be used "sparingly" while encouraging the conservation of genetic diversity. In carrying out this examination, the Services will consider available scientific evidence of the discrete population segment's importance to the taxon to which it belongs. This consideration may include, but is not limited to the following:

1. Persistence of the discrete population segment in an ecological setting unusual or unique for the taxon,
2. Evidence that loss of the discrete population segment would result in a significant gap in the range of a taxon,
3. Evidence that the discrete population segment represents the only surviving natural occurrence of a taxon that may be more abundant elsewhere as an introduced population outside its historic range, or
4. Evidence that the discrete population segment differs markedly from other populations of the species in its genetic characteristics,

Because precise circumstances are likely to vary considerably from case to case, it is not possible to describe prospectively all the classes of information that might bear on the biological and ecological importance of a discrete population segment."

This policy indicates that designation of DPS relies on determination if a population is markedly separated from others, and if it differs from others significantly. These criteria are quite subjective. Because scientific opinions will differ over these criteria, who decides what is discrete and significant is important. In the case of the ESA, it is the federal government agencies. In my opinion, DPS should not be included in the ESA definition of species. The States already do a good job managing wildlife populations and the federal government should not invoke the ESA to assume this role.

Evolutionarily significant units

Distinct population segments (DPS) for the five species of salmon and steelhead (i.e., anadromous rainbow trout) in the Pacific Ocean are called evolutionarily significant units (ESU, Federal Register 1996):

"The National Marine Fisheries Service (NMFS) has developed a Policy on the Definition of Species under the Endangered Species Act (56 FR 58612-58618; November 20, 1991). The policy applies only to species of salmonids native to the Pacific. Under this policy, a stock of Pacific salmon is considered a DPS if it represents an evolutionarily significant unit (ESU) of a biological species. A stock must satisfy two criteria to be considered an ESU:

(1) It must be substantially reproductively isolated from other conspecific population units; and

(2) It must represent an important component in the evolutionary legacy of the species."

As with the discreteness and significance criteria for DPS, the ESU criteria of "substantially reproductively isolated" and "important component in the evolutionary legacy of a species" are not definitive and determined by federal government agencies.

From a scientific standpoint, it is important to note that there are two definitions of ESU (Avise 2000:269, Cronin 2006): the one used by NMFS for the ESA described above; and one that identifies ESU as groups of populations with reciprocally monophyletic mtDNA haplotypes and significantly different nuclear allele frequencies (Mortiz 1994). Like subspecies and DPS, ESU designation is somewhat arbitrary.

The salmon and steelhead ESU designations for the ESA as defined by NMFS above are complex. For example, ESU, which are subdivisions of a species, are further subdivided into populations based on rivers (e.g., Good et al. 2005). NMFS has done an extensive amount of quantitative science identifying the distribution, and numbers of salmon and steelhead ESU. However, the appropriateness of the ESA for managing populations (i.e., ESU) rather than entire species remains a fundamental question.

Chapter 4. Polar bears, Climate Change, and the Endangered Species Act

Bullet Summary

— Polar bears were listed as a threatened species likely to become endangered with extinction under the Endangered Species Act in 2008.

— The entire polar bear species was considered likely to become threatened with extinction because of loss of sea ice in the Arctic caused by global warming and climate change.

— This was determined with models predicting declines of sea ice and its impact on polar bear populations.

— I reviewed the determination that polar bears were a threatened species by the U.S. Fish and Wildlife Service and supporting research by the U.S. Geological Survey.

— I found that the determination of polar bears as threatened with extinction was premature and not scientifically rigorous.

— In my review I suggested that the prediction of a decline of polar bear numbers should be formulated as a hypothesis to be tested with observations.

— In my review I noted that polar bears survived previous warm periods.

— Subsequent published research suggests that polar bears and other species associated with sea ice survived previous warm periods in which there was likely little or no Arctic summer sea ice.

— This suggests they could also survive ice-free summers in the future.

— My comments were dismissed or ignored by the Fish and Wildlife Service and the U.S. Geological Survey.

Chapter 4.

POLAR BEARS, CLIMATE CHANGE, AND THE ENDANGERED SPECIES ACT

Polar bears were determined to be a threatened species, likely to become endangered with extinction, under the Endangered Species Act (ESA) in a proposed rule in 2007 (Federal Register 2007). This was followed by a final rule listing the polar bear as a threatened species in 2008 (Federal Register 2008). Polar bears were thought to be threatened with extinction due to predicted loss of summer sea ice habitat caused by climate change and global warming.

I'm not an expert on climate. I do not know the extent or cause of climate change and global warming. Experts report that there has been warming, especially in the Arctic, accompanied by earlier melting in the spring, later freezing in the fall, and reduction in thickness of Arctic sea ice. The U.S. military, including the Navy and Coast Guard (U.S. Coast Guard 2013, U.S. Navy 2014) are planning accordingly, which in my view adds credibility to this issue. You can research the topic of climate change yourself if you are so inclined. My concern is with the quality of the science predicting polar bears to be threatened with extinction.

The number of polar bears worldwide has increased from about 8,000-12,000 in the 1960's (U.S. Fish and Wildlife Service 2008), to about 22,000-31,000 in 2015 (U.S. Fish and Wildlife Service 2017). The world population of polar bears is distributed among 19 subpopulations around the Arctic. Some of the subpopulations are thought to be declining, some appear to be stable, and some are doing well including the Chukchi Sea subpopulation off northwest Alaska and northeast Siberia which has about 3,000 bears. The number of bears in each subpopulation is uncertain because it is difficult to count them in the remote Arctic. Polar bears' primary food source is marine mammals, such as seals, which they hunt on the sea ice.

In this chapter I discuss the 2007 proposed rule by the U.S. Fish and Wildlife Service in which the government determined that polar bears were a threatened species (Federal Register 2007), supporting research by the U.S. Geological Survey (U.S. Geological Survey 2007), and the comments on these documents that I wrote in 2007 which are in the Appendix. Information on the status of polar bears subsequent to 2007 is available from other sources (e.g., Stirling 2011, Derocher 2012, Crockford 2019, Polar Bear Specialist Group 2019, U.S. Geological Survey 2019).

The entire species of polar bears was listed under the ESA as a threatened species. This means they are likely to become endangered with extinction in the foreseeable future (see Chapter 2 for information on the ESA). This was determined with models predicting declines of sea ice and its impact on polar bear populations. I reviewed the determination that polar bears are threatened with extinction in the proposed rule (Federal Register 2007) and supporting research (U.S. Geological Survey 2007), and found the determination that polar bears are threatened with extinction was premature and not scientifically rigorous. My comments were dismissed or ignored by the Fish and Wildlife Service and the Geological Survey. My reviews, including citations, are reproduced in the Appendix, with key points excerpted below.

In my review I summarized the issues presented in the proposed rule:

"The proposed rule can be summarized as follows:

1. Because of climate change (i.e. global warming) arctic sea ice in summer is disappearing (by melting and removal by currents).

2. This will reduce the quality and quantity of polar bear habitat. Seals are the primary prey of polar bears, and the decline in sea ice will reduce seal habitat, and access to seals by polar bears.

3. Preliminary data/observations suggest that some polar bear populations may be experiencing negative impacts at present from decreased summer sea ice. This includes possible nutritional stress, lack of access to ice because of increased open water, intra-specific predation, mortality, and decline in some populations' numbers.

4. Considering 45 years as the "foreseeable future" it is concluded that polar bears will be threatened (i.e. they will be endangered with extinction) in that time frame.

5. A review of polar bear biology, and potential impacts from melting ice and other factors (e.g., hunting, disease, oil and gas development) is presented.

6. It's not clear how well the literature is covered on predicted climate change and sea ice conditions. This deserves intensive review and assessment by experts in these fields, as it is the basis of the entire issue."

From this it can be seen that the determination that polar bears will be threatened with extinction within 45 years is based on loss of summer sea ice habitat. An important point is that the U.S. Geological Survey (2007) wrote research reports in support of the proposed rule *after* the proposed rule was written. I made this comment in my review of the U.S. Geological Survey reports (Appendix):

"It is apparent that the proposed rule and status review were premature and inadequate if additional studies were needed to assess the basic premises of current and future status of polar bear populations."

The Fish and Wildlife Service and U.S. Geological Survey are both agencies in the Department of the Interior, and the U.S. Geological Survey did research to support the Fish and Wildlife Service proposed rule. The title of the U.S. Geological Survey report is "USGS Science Strategy to Support U.S. Fish and Wildlife Service Polar Bear Listing Decision". This is relevant to my point in Chapter 1 about the government control and monopoly of the science information used for the ESA. I made this point in my review of the Proposed Rule:

"The entire system of review of science in the ESA process is closed...The agency basically has the role of author, editor, and reviewer for their own documents...courts give deference to agencies in assessing science and management information."

Another of my comments on the proposed rule to list the polar bear as a species threatened with extinction was regarding polar bears' survival during previous warm periods:

"Predictions of future polar bear population status should also consider that polar bears survived previous warming periods, including periods 8,000-11,000 years ago and 1000 years ago (Buck 2007). This is noted in the Proposed Rule (page 72) but no insights are offered regarding the predicted warming of concern to extant populations. It is well known that many species went extinct about 12,000 years ago (termed the megafaunal extinction) but this didn't include polar bears. Interestingly, cheetahs also escaped the megafaunal extinctions of 12,000 years ago but suffered a severe loss of genetic variation, presumably from a population bottleneck at that time (O'Brien and Johnson 2005). Genetic variation appears not to be reduced in polar bears compared to their ancestral species (brown bears, Paetkau et al. 1997, 1998, 1999, Cronin et al. 2005, 2006). This is not hard evidence there was not a population bottleneck of polar bears during previous warming periods, but it is suggestive that there were not population reductions severe enough to reduce genetic variation as in cheetahs. FWS willingness to predict future polar bear population status should be complemented with a backward look at past populations and habitat conditions."

Subsequent to submitting these comments, colleagues and I published scientific papers supporting the premise that polar bears survived previous warm periods. One interesting finding is that fossil and genetic (DNA) evidence suggests that polar bears have existed as a species for at least 125,000 years and maybe much longer (Cronin et al. 2014 and references therein). Cronin and Cronin (2015) noted that DNA and fossil evidence indicates polar bears and their prey, such as ringed seals and walruses, survived for the last 125,000 years and probably longer. This means they experienced the extreme climates of glacial and interglacial periods, including summers that were partially or completely without Arctic sea-ice.

Loss of summer sea ice could have serious impacts on polar bears and marine mammals. However, polar bears and other species associated with sea ice survived previous periods that likely had ice-free summers as described by Cronin and Cronin (2015). This suggests they could also survive ice-free summers in the future. It is important to note that the predicted loss of Arctic sea ice is for the summer and the multi-year ice pack. The Arctic Ocean is expected to continue freezing in winter. These observations may be useful in the current assessments of impacts from recent declines in sea ice, as well as reassessment of the rigor of the 2008 determination that polar bears are threatened and likely to become endangered with extinction.

I also pointed out in my comments on the proposed rule that the standard scientific practice of hypothesis testing should be used, and the prediction of a decline of polar bear numbers should be formulated as a hypothesis to be tested with observations:

> "A broad assessment of the Proposed Rule shows that there is no presentation or testing of hypotheses, despite the fact that the relevant assessment of polar bears and their habitat is predictive in nature. Simple assumptions that warming and ice disappearance will continue and that polar bears will decline across their entire range to the point of near extinction are made. It is reasonable to conclude that if sea ice declines significantly or entirely, polar bear populations will decline. However no quantitative analysis or models of population numbers (or prey abundance) are made. The analysis is speculative."

It has now been 12 years since the polar bear ESA listing was proposed in 2007. That is 26.7% of the 45-year foreseeable future used in the ESA listing. There should be enough data now to test a hypothesis regarding polar bear population numbers for this time period. I am not aware if such an effort by the government is underway.

Chapter 5 Forests, Spotted Owls, and the Endangered Species Act
Bullet Summary

— Conflict over timber harvest and potential impacts to wildlife in the western United States and Alaska was intense in the 1990's and continues.

— Environmentalists claimed that logging and timber harvest negatively impact wildlife.

— Others argued that forests can be managed for multiple-use to produce timber and protect wildlife.

— The focus of the conflict was in the Pacific Northwest where the Endangered Species Act (ESA) listing of the northern spotted owl resulted in major decreases of timber harvest.

— This decrease in timber harvest devastated the timber industry and communities reliant on it.

— The northern spotted owl is a subspecies of the spotted owl species.

— The California spotted owl and Mexican spotted owl are also subspecies of the spotted owl species.

— The ranges of the northern spotted owl and California spotted owl overlap.

— There is some movement and interbreeding of owls among the three spotted owl subspecies.

— Subspecies in general, including the northern spotted owl, are not scientifically definitive.

— The northern spotted owl designation as an endangered subspecies is an example of the selective use of equivocal science with the ESA.

Chapter 5

FORESTS, SPOTTED OWLS, AND THE ENDANGERED SPECIES ACT

osworth and Brown (2007) declared in an article that "The timber wars are over". The timber wars refer to conflicts over logging and wildlife that were particularly intense in the 1990s. Environmentalists and wildlife advocates claimed that logging and timber harvest negatively impacted wildlife, and the timber industry and others argued that forests could be managed for multiple-use to produce timber and protect wildlife. Much of the focus of the conflict was in the Pacific Northwest States of Washington, Oregon, and northern California where the Endangered Species Act (ESA) listing of the northern spotted owl resulted in the devastation of the timber industry and communities reliant on it. Alston Chase (1995) describes this episode of American history in great detail. Conflicts over timber harvest and wildlife impacts also occurred in southeast Alaska and other areas.

I think that Dale Bosworth, a former chief of the U.S. Forest Service and Hutch Brown a Forest Service analyst (Bosworth and Brown 2007), were wrong. The timber wars are not over. Bosworth and Brown wisely suggest replacing conflict with collaborative community-based forest stewardship, but the ESA continues to be used to stop timber harvest, and also oil and gas extraction, mining, livestock grazing, and other land uses. Perhaps many people accept the environmental community's success in stopping natural resource use with the ESA and other regulations. I don't, because I know the ESA is often implemented with the selective use of equivocal science while claiming it is using the best available science.

I also know that economic hardship and the loss of jobs in communities that depended on the timber industry had serious impacts on my fellow citizens. A former Chief of the Forest Service, Jack Ward Thomas, also had concerns about our fellow citizens because of the great reduction of timber harvest in the Pacific Northwest (Thomas 2005). He was haunted by the thought of thousands of people hurt by the loss of timber industry jobs because of the spotted owl ESA listing, but he had to deal with the biology of the situation.

Thomas also noted that the ESA requires the preservation of the old-growth forest ecosystem and this drove the policy of stopping timber harvest. This was primarily because of the ESA listing of one subspecies: the northern spotted owl.

The biology of the situation was that the numbers of northern spotted owls were declining and this was thought to be because of loss of old growth forest habitat resulting from logging. Competition with barred owls is also a threat to northern spotted owls. But was that really the biology relevant to the ESA listing? I don't think it was. The ESA requires determination of whether a species is threatened or endangered with extinction. Perhaps the northern spotted owl was close to extinction because of logging, but this was uncertain (Chase 1995). The ESA also requires that the target of a listing petition be a species. As discussed in Chapter 3, this can be a species, subspecies, or distinct population segment (DPS). Subspecies and DPS are not scientifically definitive. The northern spotted owl is a subspecies, with the inherent uncertainty of the category.

The Northern Spotted Owl and Wolf subspecies

The northern spotted owl is perhaps the best known subspecies listed under the ESA. This ESA listing caused major damage to the timber industry in the Pacific Northwest. See Alston Chase's (1995) book *In a Dark Wood* for a thorough and scholarly description of this major environmental conflict.

Was the northern spotted owl ESA listing justified? I think it was not. Part of the justification was determination of whether the northern spotted owl is indeed a subspecies. The general subjectivity and lack of scientific consensus about subspecies in general (see Chapter 3) indicates that the northern spotted owl subspecies designation is not definitive. The relationship of the northern spotted owl with other spotted owl subspecies also indicates an uncertain subspecies status.

There are three subspecies of spotted owl recognized: the northern spotted owl, the California spotted owl, and the Mexican spotted owl. The northern spotted owl range is in southwest British Columbia, western Washington and Oregon, and northern California. The northern spotted owl range overlaps that of the California spotted owl which is in the southern Cascade Mountains, the western Sierra Nevada Mountains, and mountains in southern California. The Mexican spotted owl range is in parts of Utah, Colorado, Arizona, New Mexico, west Texas, and Mexico. The northern spotted owl and the Mexican spotted owl are listed under the ESA, and the California spotted owl is not listed but has been petitioned for ESA listing with a decision due in 2019. Considering that the California spotted owl has not been considered endangered, simple logic indicates that the entire spotted owl species is not endangered. Allowing subspecies to be listed under the ESA resulted in the northern spotted owl situation.

Genetic studies are used to support the spotted owl subspecies designations but they are not definitive. First, the general subjectivity of the subspecies category makes all subspecies designations uncertain. Second, the genetic data used to support the subspecies designations is equivocal. There is movement and interbreeding of owls among all three spotted owl subspecies. One genetic study did not clearly differentiate northern spotted owls and California spotted owls (Haig et al. 2001). Other research supports the spotted owl subspecies (Barrowclough et al. 1999, 2005, Haig et al. 2004, Funk et al. 2008) but these studies do not show complete genetic differentiation. There is genetic differentiation of allele frequencies and mitochondrial DNA (mtDNA) haplotype frequencies, but not clear phylogenetic differentiation. Importantly, introgression (i.e., interbreeding with resulting gene flow) was shown between northern spotted owls and California spotted owls in northern California and southern Oregon (Funk et al. 2008). There is also movement and introgression of Mexican spotted owls into the range of northern spotted owls. This indicates that the ranges of the subspecies overlap and there is interbreeding among them. This pattern does not support subspecies designations in my view. As noted, subspecies designations are subjective, so designating or not designating spotted owl subspecies is a matter of opinion.

As in other cases, the northern spotted owl ESA listing was based on equivocal science. In my opinion, ESA listings such as the northern spotted owl used indefinite subspecies designations to stop natural resource development, in this case timber harvest.

Another subspecies considered for ESA listing is the wolf in southeast Alaska. These wolves are considered by some biologists to be the Alexander Archipelago wolf subspecies (MacDonald and Cook 2009, Weckworth et al. 2015). The Alexander Archipelago wolf subspecies occurs in southeast Alaska and coastal British Columbia. Other biologists consider the wolf in southeast Alaska to be part of the plains wolf subspecies (Chambers et al. 2012). The original range of the plains wolf covered much of the western United States.

The Alexander Archipelago wolf was petitioned for ESA listing as a threatened or endangered subspecies. The government acknowledged the uncertainty of the subspecies designation but assumed the Alexander Archipelago wolf is a valid subspecies for the purpose of the ESA finding. However, it was found not warranted for listing because it was not threatened or endangered with extinction (Federal Register 2016).

Morphological and genetic analyses indicate the Alexander Archipelago wolf is not distinct enough from other wolves to warrant subspecies status. This led Chambers et al. (2012) to consider wolves in southeast Alaska to be the plains wolf subspecies. Genetic analyses show there is as much variation among the wolves in different parts of southeast Alaska as there is between wolves in southeast Alaska and other areas. There is genetic differentiation of wolves in southeast Alaska from those in other areas but it is not remarkably large. These data were considered insufficient to support a subspecies designation (Cronin et al. 2015a, 2015b). Others think the Alexander Archipelago wolf is genetically distinct enough to warrant subspecies status (Weckworth et al. 2015). This is an example of differing opinions over a subspecies designation considering the same data.

The Alexander Archipelago wolf was found not warranted for ESA listing. If it had been listed, it likely would have been used to support lawsuits to stop timber harvest on private land and State of Alaska land in addition to the Tongass National Forest. This is because the ESA applies on federal land, and also on private and state land.

Even without an ESA listing of the wolf, timber harvest was already greatly reduced on the Tongass National Forest in southeast Alaska because of concerns over wildlife habitat. Regarding this issue, biologists with The Wildlife Society (TWS) proposed science-based conservation advocacy in southeast Alaska because of concerns about impacts of logging on wildlife (Kirchoff et al. 1995). The goal was to use science to support management objectives of minimizing impacts on wildlife by reducing logging of old growth forest. Because many of the TWS members work in government agencies, the issue of advocacy by government agency employees is a serious consideration. In response to the TWS' proposed advocacy, I noted (Cronin 1996) that it is inappropriate for government agency employees to advocate for personal management objectives and they should try to achieve the management objectives of the public.

BULLET SUMMARY CHAPTERS 1-5: THE ENDANGERED SPECIES ACT PROBLEMS AND SOLUTIONS

The Endangered Species Act (ESA) has problems with science and policy.

1. Science problems with the ESA are:
   a. The ESA species definition includes subspecies, distinct population segments-DPS, and evolutionarily significant units-ESU which are not scientifically definitive categories.
   b. The risk of extinction expressed as being threatened or endangered, is not scientifically definitive.
   c. What constitutes a significant portion of a species range and the foreseeable future are not scientifically definitive.
   d. Federal government agencies make all of the science determinations for the ESA.

The solution is to:
   a. Change the ESA to apply only to full biological species.
   b. Separate the regulatory and science functions of the ESA so they are not in one agency.
   c. The Fish and Wildlife Service and National Marine Fisheries Service can implement the ESA using science determinations made by others.
   d. The science relevant to the ESA can be done outside government under contract to universities and private sector science companies and institutions, and include rigorous peer review.

2. Policy problems with the ESA are:
   a. The ESA applies on private property.
   b. The federal government takes fish and wildlife management authority from the States and gets deference in court.
   c. The ESA causes a burden of extensive litigation on governments, industries and agriculture, and private landowners.

The solution is to:
   a. Change the ESA so it does not apply on private property. Provide voluntary incentives for landowners to protect and enhance habitat for wildlife.
   b. Change the ESA to make explicit that States have exclusive jurisdiction over fish and wildlife populations, including subspecies, DPS, and ESU.
   c. Change the ESA so the States and federal government have equal co-authority on ESA decisions. Provide incentives for States to protect and enhance habitat for wildlife.
   d. End the granting of deference to federal agencies in court, and change or repeal the Equal Access to Justice Act.

Chapter 6 The North Slope of Alaska: Oil Fields and Caribou
Bullet Summary

— Oil and gas fields have been developed on the North Slope of Alaska since the discovery of a major oil reservoir at Prudhoe Bay in 1968 and completion of the Trans-Alaska Pipeline in 1977.

— Concerns over environmental impacts, particularly to fish and wildlife, have been paramount on the North Slope resulting in extensive research and monitoring.

— Impacts of the oil fields on caribou have been a primary concern.

— The Central Arctic caribou herd uses habitats in and around the North Slope oil fields for calving and post-calving summer range.

— Disturbing and displacing caribou from calving areas because of oil field activity has been postulated as a negative impact.

— Some studies show displacement of calving caribou from oil field infrastructure and other studies do not.

— The extent of this impact has been debated, particularly when the studies at Prudhoe Bay are used to predict impacts of oil development in the Arctic National Wildlife Refuge (ANWR).

— The Central Arctic caribou herd grew from 5,000 animals in the late 1970s when the oil fields were first developed to 68,000 in 2010, and declined to 28,000 in 2017.

— Studies of population genetics and population dynamics show that inter-herd movements substantially affect the numbers of caribou in each herd, and data do not support the hypothesis that oil field impacts have caused a population decline.

— The impact of oil fields on the number of caribou in the Central Arctic herd is probably small compared to other factors including winter severity and habitat, immigration into and emigration out of the herd, and calf recruitment.

— Oil field impacts continue to be debated and are used to oppose development in ANWR

— Government reports do not adequately consider published literature on caribou that was sponsored by the oil industry.

Chapter 6

THE NORTH SLOPE OF ALASKA: OIL FIELDS AND CARIBOU

O il and gas fields have been developed on the North Slope of the Brooks Mountain Range in northern Alaska since the discovery of a major oil reservoir at Prudhoe Bay in 1968 and completion of the Trans-Alaska Pipeline in 1977 (Truett and Johnson 2000). In addition to the Prudhoe Bay field, several other fields have been developed onshore and offshore on the North Slope. Energy production in general and the Trans-Alaska Pipeline (TAPS) in particular are important to U.S. national security. Alaska's North Slope is in the Arctic, and the Arctic is also increasingly important to national security (U.S. Coast Guard 2013, U.S. Navy 2014). See my essay in the Appendix about oil and national security (Cronin 2013).

Since the beginning of oil field development on the North Slope, concerns over environmental impacts, particularly to fish and wildlife, have been paramount. This has resulted in extensive research (e.g., Truett and Johnson 2000, Trans-Alaska Pipeline System Owners 2001, Douglas et al. 2002, National Research Council 2003, Pearce et al. 2018). Impacts of the oil fields on caribou have been a primary concern.

The Central Arctic caribou herd occurs in and around the North Slope oil fields in the summer, before migrating south to winter ranges. The Porcupine River caribou herd occurs to the east of Prudhoe Bay, and its calves are often born in the Arctic National Wildlife Refuge (ANWR). Two other Arctic Alaska herds, the Teshekpuk Lake caribou herd and Western Arctic caribou herd, occur to the west of Prudhoe Bay. The four herds' ranges overlap and numbers of caribou have fluctuated over time (Tables 1-4 and Figures 1-4).

It has been postulated that disturbing and displacing caribou from calving areas because of oil field activity is an impact that negatively affects herd productivity (i.e., calf recruitment into the population). The extent of this impact has been debated, particularly when the studies at Prudhoe Bay are used to predict impacts of potential oil development in ANWR. This resulted in considerable controversy. For example, in 2001 the U.S. Fish and Wildlife Service (which is in the Department of the Interior) provided information to the Secretary of the Interior regarding caribou and potential oil development in ANWR. The Secretary synthesized the information from the Fish and Wildlife Service with information from industry-sponsored published papers on caribou and provided this to the U.S. Senate. Fish and Wildlife Service employees criticized the Secretary for her assessment and the information she provided to Congress (Grunwald 2001). I responded to their criticisms (Cronin 2001). I believe the Fish and Wildlife Service employees were opposed to oil development in ANWR and provided selected information to the Secretary, ignoring science that did not support an anti-development agenda. The Senate subsequently failed to pass a bill to allow oil development in ANWR (Rosenbaum 2002).

There is a large literature on caribou and oil fields, which I don't thoroughly review here because it would take a large volume. In this chapter I summarize important aspects of this issue including government reports (National Research Council 2003, Bureau of Land Management 2018, Pearce et al. 2018). These government reports are important because they have been, and will be, considered as authoritative sources by Congress, the Executive Branch, the courts, and the news media.

In my opinion, government agencies and environmental groups have selectively used information, with inadequate consideration of oil industry sponsored research, to present potential negative impacts of oil development in ANWR. I believe their primary objectives are preserving ANWR as undeveloped wilderness, generally stopping resource development, and generating funding. I think that the impacts on wildlife with emotional appeal to the public are exaggerated through the selective use of information to achieve these objectives.

I presented detailed information on these issues in written testimony to the U.S. Senate Energy and Natural Resources Committee (Cronin 2017), some of which is excerpted here.

Excerpt from Testimony to the U.S. Senate Committee on Energy and Natural Resources (Cronin 2017).

Note: The entire testimony, including citations, is available at the website in the Cronin (2017) citation. The tables and figures here (Tables 1-4, Figures 1-4) have been updated with census data from 2017, subsequent to submission of the written testimony in November 2017. See the written testimony for the citations in this excerpt.

"Caribou calving (i.e., cows giving birth to calves) in the Alaska North Slope oil fields has been studied extensively. On the North Slope, caribou cows give birth to calves in late May and the first two or three weeks of June. Some studies reported statistically significant lower density of calving caribou within 1 kilometer (km) of oilfield roads and facilities than in areas farther from the roads and facilities (Cameron et al. 1992). Non-statistically significant lower densities of calving caribou were also observed within 4 km of roads and facilities (Dau and Cameron 1986, Cameron et al. 1992). These observations were interpreted as displacement from, and avoidance of, oil field roads and facilities. However, these studies showed that some calving occurs within 1 km and within 4 km of roads and facilities. That is, there was not complete absence of calving caribou within 1 km or within 4 km of the oilfield roads and facilities. In the study reporting displacement and avoidance of the area within 4 km of an oil field road, 44.4% of the calves occurred from 0 to 4 km of the road, and 55.6% occurred from 4 km to 6 km of the road, so there was not complete avoidance of the areas within 4 km of the roads and facilities (Cameron et al. 1992, Noel et al. 2004). A replicate study actually found higher densities of calves within 1 km of the roads than farther away from the roads (Noel et al. 2004, 2006a).

These studies indicate that lower calf density sometimes occurs, and sometimes does not occur, within 1 km of roads and facilities. Factors including habitat (e.g., types of vegetation), timing of snow melt in the spring (persistent snow influences how far north caribou move as they migrate from their winter ranges that are south of the oil fields), and perhaps most importantly, habituation (caribou learn over time that vehicles, roads, and buildings are not a threat) influence where caribou calve (Haskell and Ballard 2008, Haskell et al. 2006).

General calving areas also shift over time. This has been reported for caribou herds with no oil development within their ranges and is a natural occurrence (Noel et al. 2006a and references therein). A shift of calving concentration to the south of the North Slope oil fields has been attributed to oil field expansion on the North Slope (Joly et al. 2006 and references therein). However, this is not definitive because the same factors affecting individual caribou responses to roads and facilities (e.g., timing of spring snowmelt, habitat, habituation, predators and human hunting) also affect the general areas in which caribou calve (Haskell and Ballard 2008, Haskell et al. 2006).

After the calving period, caribou do not avoid the oil fields and travel through them regularly (Pollard et al. 1996a, Cronin et al. 1998a, Noel et al. 1998, 2006b). Pipelines and roads can block or deflect caribou movements, but elevating pipelines above the ground, separating pipelines and roads, and other measures minimize this impact. There is intense mosquito and fly harassment of caribou during the summer, and caribou will travel to the coast where there is more wind, and also often congregate on the oil field roads and gravel pads and in the shade of buildings and pipelines to escape the insects (Pollard et al. 1996b, Noel et al. 1998). There is no hunting allowed in the oil fields, which gives caribou protection. Proper design and operation of oil fields has actually enhanced caribou habitat in important ways.

It has been hypothesized that the oil fields have negatively affected caribou reproduction and recruitment (i.e., calf birth and survival) and the number of animals in the Central Arctic herd (National Research Council 2003 and references therein). In particular, a decline of the Central Arctic herd of 5,344 animals between 1992 and 1995 (Table 1 and Figure 1) was partially attributed to oilfield impacts (National Research Council 2003 and references therein). Several lines of data do not support this hypothesis (Cronin et al. 1997, 1998b, 2000, Haskell and Ballard 2004, Noel et al. 2006). First, note that the decline in the Central Arctic herd was small relative to the overall changes in the four Arctic Alaska herd numbers over time (Tables 1-4, Figures 1-4, note these tables and figures are updated with 2017 data since the Senate testimony). Increases or decreases of the magnitude seen in the Central Arctic herd between 1992 and 1995 are well within the natural variation of herd numbers. Second, note that the Teshekpuk herd also declined by a similar amount (8,951 animals) in the same time period (between

1993 and 1995) without oil fields in its range (Table 3, Figure 3). Third, consider that the caribou of the Central Arctic herd spend about 10 months of the year in ranges away from the oil fields where many other factors affect their survival and fitness (Cronin et al. 2000). It is becoming apparent from studies of population genetics and population dynamics that inter-herd movements substantially affect the numbers of caribou in each herd, and data do not support the hypothesis that oil field impacts have caused a population decline (Cronin et al. 1997, 1998b 2000, 2003, 2005, ADFG 2016).

The Central Arctic caribou herd has grown since the oil fields and Trans-Alaska Pipeline were developed. There have been fluctuations in numbers (Table 1, Figure 1, note this table and figure are updated with 2017 data since the Senate testimony), but neither the increases nor decreases can be attributed to impacts of the oil fields as noted by the Alaska Department of Fish and Game:

'The impact of oil infrastructure on CAH (Central Arctic herd) has also been considered, but *is not thought to be contributing to the decline* since the herd grew substantially during peak oil development.' (ADFG 2016, my italics).

In contrast, there is empirical evidence that emigration contributed to a decline in the numbers of caribou in the Central Arctic herd between 2013 and 2016:

'From 2013 to 2015, extensive mixing occurred between the CAH (Central Arctic herd), Porcupine, and Teshekpuk herds after calving and during the winter. Several thousand caribou left CAH and joined other herds.' (ADFG 2016).

This is seen in the herd numbers (Table 1, note this table is updated with 2017 data since the Senate testimony). The number of caribou in the Central Arctic herd census in 2013 was actually about 70,000 but there were animals from the Porcupine caribou herd present, so the herd estimate was adjusted downward to 50,753 (Parrett et al. 2014, Lenart 2015). The decline between 2013 and 2016 was likely due in part to high female mortality and emigration (Bohrer 2017, Cotten 2016).

The numbers of caribou in the Central Arctic herd reflect habitat, winter severity, inter-herd emigration and immigration, population density, and other factors described in the published literature. The hypothesis that changes in herd numbers are due to oil field impacts has not been supported considering all of the available data."

Comments on government reports

The National Research Council (NRC) conducts studies on science issues important to the nation. One such study was completed in 2003: Cumulative Environmental Effects of Oil and Gas Activities on Alaska's North Slope (National Research Council 2003). I reviewed the chapter on caribou in this report and found it did not consider all of the information available regarding caribou and the North Slope oil fields. It lacks proper review of published scientific papers, relies heavily on unpublished material, and did not thoroughly consider nor cite industry-funded published papers.

It is important to note that the NRC committee concluded that the oil fields have had considerable impacts on caribou, when another federal study at about the same time, the Environmental Impact Statement (EIS) for renewal of the Trans-Alaska Pipeline right-of-way (Bureau of Land Management 2002), did not. This EIS was prepared by the highly-experienced Department of Energy (DOE) Argonne National

Laboratory for the Bureau of Land Management. Unlike the NRC committee, the Department of Energy and Bureau of Land Management refrained from making uncertain cause-effect relationships.

In addition, census data for the Central Arctic herd, showing an increase from 29,519 in 2000 to 34,211 animals in 2002 was absent from the NRC report. Extensive data in published papers (Maki 1992, Pollard et al. 1996a, 1996b, Cronin et al. 1997, 1998a, 1998b, 2000, 2001, Noel et al. 1998, Ballard et al. 2000) were also not adequately considered in the NRC report. These published papers show large numbers of caribou using oil field habitats, quantification of population density in developed and undeveloped areas, the potential for immigration and emigration to affect herd numbers, high calf production in the oil field areas, and continued growth of the herd despite oil field development.

The National Research Council (2003) chapter on caribou is speculative in suggesting cause and effect relationships from non-definitive studies. The most prominent of these is the linkage of studies showing some displacement of caribou from oil field structures, and a lower parturition rate in the oil field areas. The NRC Committee considered these and other data to infer that Central Arctic herd cows (i.e., adult females) in contact with oil development from 1988 through 2001 had lower reproductive success than for cows in other areas, and that this contributed to a reduction in herd productivity.

This cause-effect speculation was made despite the high calf production in the oil field areas in most years, and the steady growth of the herd from 5,000 before oil development to about 34,000 in 2002 when the NRC report was written (Table 1 and Figure 1). Subsequent growth of the Central Arctic herd to 68,000 caribou in 2010 also indicates it is unlikely that the oil fields were responsible for population declines. The data on this topic are complex, and warrant detailed review (see Cronin et al. 2000, 2001).

The U.S. Geological Survey (USGS) Ecosystems Mission Area is the biological research arm of the Department of the Interior. The USGS published wildlife research summaries in 2002 and 2018 (Douglas et al. 2002, Pearce et al. 2018) for the coastal plain of the Arctic National Wildlife Refuge (ANWR).

The USGS report (Pearce et al. 2018) does not adequately cover the scientific literature on caribou. Several peer-reviewed, published, scientific papers on caribou on Alaska's North Slope were not included in the USGS report, resulting in an incomplete and scientifically deficient assessment. This literature reports the results of research funded by the oil industry, and is legitimate, published science. This includes papers described and cited in my testimony submitted to the Senate Energy and Natural Resources Committee (Cronin 2017) that would have been available to the authors of the USGS report. For example, a study on calving caribou distribution (Noel et al. 2004, 2006) was not cited in the USGS report, although a critique of this study *was* cited (Joly et al. 2006). Research on caribou habituation to the oil fields was also not adequately covered (Haskell and Ballard 2004, 2008, Haskell et al. 2006). Also, some points made in the USGS report regarding potential factors impacting caribou populations did not properly cite the original sources. These factors include population density and interherd movements (Cronin et al. 1997, 1998b, 2000, 2005, and references in these papers).

The papers not referenced in the USGS report present data and analyses relevant to assessing impacts of the Alaska North Slope oil fields on wildlife. These papers demonstrate that several factors other than oil field impacts need to be considered in assessing caribou populations. The papers also show that there have been limited negative impacts of the oil fields on caribou populations and document positive impacts of the oil fields.

The selective use of North Slope oil field biological literature has occurred previously. Like the 2018 USGS report, a previous assessment of the North Slope oil fields by the USGS (Douglas et al. 2002) also neglected much relevant literature. The 2018 USGS report needs to be revised with proper consideration of the scientific literature. This is particularly important because regulatory documents such as the National Environmental Policy Act (NEPA) Environmental Impact Assessments of the coastal plain of ANWR, will rely on such reports. Omission of relevant literature can lead to biases in Environmental Impact Statements and unnecessary controversy (see Cronin 2001, Grunwald 2001).

The result of selective use of science is seen in the December 2018 Coastal Plain Oil and Gas Leasing Program draft Environmental Impact Statement (Bureau of Land Management 2018) which did not adequately cite relevant industry funded published papers. One important example is the prediction of displacement of calving caribou 2.49 miles from oil field infrastructure (pages 3-112 and 3-114 of the draft Environmental Impact Statement), despite published literature that indicates this is not warranted (Noel et al. 2004, 2006). This indicates the importance of using all relevant information when assessing science and its application in policy and management.

Table 1. Numbers of caribou in the Central Arctic caribou herd, 1975 to 2017 (from Alaska Department of Fish and Game censuses).

Year	Number of Caribou
1975	5000
1978	5000
1980	5000
1981	8537
1983	12905
1985	15000
1989	18000
1991	19046
1992	23444
1995	18100
1997	18824
2000	29519
2002	34211
2008	66666
2010	68442
2013	50753
2016	22630
2017	28051

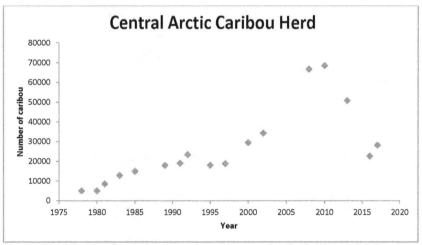

Figure 1. Numbers of caribou in the Alaska Central Arctic caribou herd, 1975 to 2017 (from Alaska Department of Fish and Game censuses).

Table 2. Numbers of caribou in the Alaska Porcupine caribou herd, 1961 to 2017 (from Alaska Department of Fish and Game censuses).

Year	Number of Caribou
1961	110000
1964	140000
1972	99959
1977	105000
1979	105683
1982	125174
1983	135284
1984	149000
1985	165000
1986	182500
1987	165000
1989	178000
1992	160000
1994	152000
1998	129000
2001	123000
2010	169000
2013	197000
2017	218457

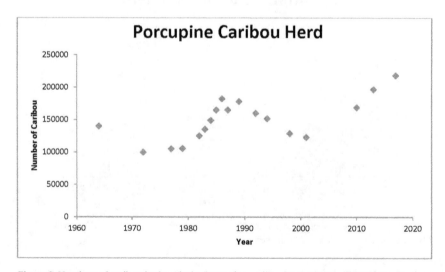

Figure 2. Numbers of caribou in the Alaska Porcupine caribou herd, 1961 to 2017 (from Alaska Department of Fish and Game censuses).

Table 3. Numbers of caribou in the Alaska Western Arctic caribou herd, 1950 to 2017 (from Alaska Department of Fish and Game censuses).

Year	Number of Caribou
1950	238000
1961	156000
1962	187500
1964	300000
1970	242000
1975	100000
1976	75000
1977	75000
1978	107000
1980	138000
1982	217863
1986	229000
1988	343000
1990	417000
1993	478822
1996	463000
1999	444597
2003	490000
2007	381501
2009	355828
2011	324963
2013	234757
2016	201000
2017	259069

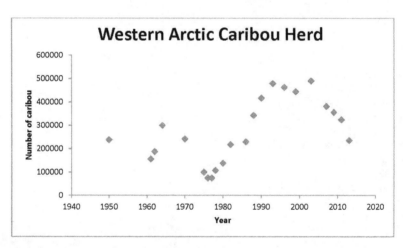

Figure 3. Numbers of caribou in the Alaska Western Arctic caribou herd, 1950 to 2017 (from Alaska Department of Fish and Game censuses).

Table 4. Numbers of caribou in the Alaska Teshekpuk caribou herd, 1978 to 2017 (from Alaska Department of Fish and Game censuses.

Year	Number of caribou
1978	3550
1981	3009
1984	18292
1985	13406
1989	19724
1993	41800
1995	32839
1999	28627
2002	51783
2008	68932
2011	55704
2013	39172
2015	41500
2017	56255

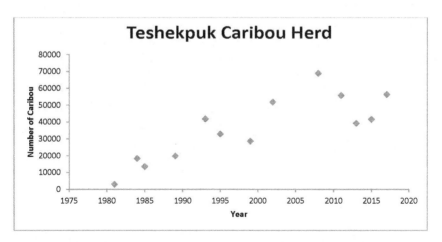

Figure 4. Numbers of caribou in the Alaska Teshekpuk caribou herd, 1978 to 2017 (from Alaska Department of Fish and Game censuses.

PART 2. CONFLICTS IN SCIENCE AFFECTING SOCIETY

Chapter 7 Biology and War
Bullet Summary

— War has been a constant worldwide characteristic of mankind throughout history.

— War is defined as open and declared armed hostile conflict between states or nations.

— War is considered by military scholars as the continuation of policy by other means.

— War is a part of human biology and there are biological causes of war.

— Men have been the combatants in war throughout history.

— The U.S. government instituted full integration of women into all Military Occupational Specialties (MOS) including combat MOS in 2015.

— This government policy was justified because of a perceived need to take full advantage of every individual who can meet military standards.

— Studies showing poorer performance of male-female integrated units than all-male units were dismissed by the government without legitimate justification.

— There is a vast scientific literature on the differences between men and women and why such differences exist.

— Science indicates that the historical pattern of men fighting men and the exclusion of women as combatants in war is likely rooted in basic differences of the sexes that developed as adaptations that enhanced fitness in nature.

— This science was apparently not considered or was dismissed by the government in the decision to integrate women into combat MOS.

— Scientific knowledge can contribute to an understanding of the historical pattern of men fighting men in war, and inform future policy decisions on the roles of the sexes in the U.S. military.

— Changing a historical pattern by allowing women to be integrated with men and fight in war entails risks to military effectiveness and the basic structure and fitness of human populations.

— The risks associated with full integration of the sexes in the military have not been adequately assessed and should be reconsidered in light of biology.

Chapter 7

BIOLOGY AND WAR

W hat is war? Webster's Dictionary states that war is "open and declared armed hostile conflict between states or nations." Military scholars often address this question by citing Clausewitz (1832) that war is "the continuation of policy by other means." Keegan (1993) and Murray (2017) provide useful insights regarding Clausewitz's assessment of war.

Clausewitz's definition implies a proximate cause for war (i.e., to achieve policy objectives). Another proximate cause of war has been identified: women. When asked why men go to war, T. E. Lawrence said "Because the women are watching", and Darwin (1871) quoted Horace: "*nam fuit ante Helenam mulier teterrima belli causa*" (for even before Helen [of Troy] a woman was a most hideous cause of war.). Women are also a cause of war among primitive tribes (e.g., Wilson 1978:115, Keegan 1993:94-115). It is debatable whether women are a general cause of war, but men fighting men to impress or gain access to women is ultimately for reproduction, and hence biological.

Modern biology can provide additional insights into the ultimate causes of war. Why humans wage war is not as obvious as other fundamental human behaviors (e.g., why humans eat, sleep, breathe, and mate), but war has been constant in all human cultures (Wilson 2012). Scientists have tried to address the ultimate causes of war considering human behaviors (e.g., aggression, territoriality) and their possible genetic and cultural roots (e.g., Shaw and Wong 1989, Vayda 1974, Wilson 2012). These assessments have been largely outside the fields of military science and history. However, a recent integration of biological and military insights on war is "The Nature of War Theory", in which Colonel Paul Olsen (U.S. Army retired) describes war in terms of evolutionary biology and Darwinian natural selection (Olsen 2011a, 2011b). This theory relates to a basic biological question: Is war innate? Or in today's slang, is war "in our DNA"? In other words, what is the biological basis of war and are biological imperatives driving decisions about when and how to wage war?

When considering complex traits, such as human behavior and aggression, it is important to guard against unwarranted cause-effect speculation regarding the genetic basis of such traits. Human behaviors, as with most phenotypic traits, are determined by a complex interaction of genetics and environment. Advances in genetics allow some resolution of genetic and environmental influences on behavior (e.g., Mukherjee 2016), but it is still a very complex and difficult field of study.

Nevertheless, war is an integral part of the human condition and its ultimate causes are suitable for scientific investigation. This has been done by scholars over the last 150 years since Darwin's (1871) "Law of Battle" regarding fighting within and between groups of male animals, including humans. Comparative studies of animals have contributed to these efforts, examining for example, territoriality and other behaviors (e.g., Wilson 2012, Wrangham and Glowacki 2012). The development of the science of sociobiology (Wilson 1975) allows a comparative approach to aggression in man and animals and its relevance to war. Sociobiology is a field of biology about social behavior and evolution with relevance to anthropology, archaeology, ecology, genetics, and psychology.

I started thinking seriously about war from a biological perspective following the announcement in 2015 of the U.S. military fully integrating women into all military occupational specialties (MOS) including combat MOS (Carter 2015, Kamarck 2015). Previously women were excluded from combat MOS, and historically women were not combatants in war (van Creveld 1991, Keegan 1993).

I must qualify here that women have the capacity to fight in combat and our military has many skilled and courageous women who have fought, or are willing to fight, alongside men. That is not the issue I am addressing. I am discussing the science behind the historical pattern of men fighting men in war. I hope the reader understands this, and that I respect the bravery and dedication of women in the U.S. military.

Here I review the issue of the sexes in the military and war. This is consistent with the theme of this book because basic science and history were summarily dismissed by the government in the decision to fully integrate women into all military MOS. Also, integration of the sexes in the military is a feminist agenda (Mitchell 1998, Gutmann 2000, van Creveld 2001, Maginnis 2013) and the feminist goal of equality for women is associated with socialism/communism (e.g., Mojab 2015, Smith 2015, Studer, 2015), as with aspects of environmentalism and religion noted in other chapters of this book.

Sex and War: Integration of women in the military: biological considerations

Solzhenitsyn (1974) expressed the sad result of government-imposed equality for women under communism in the Soviet Union in his *Letter to the Soviet Leaders*. He noted that the Soviets boasted about women's equality, but he lamented with shame and compassion the sight of their women engaged in hard labor such as carrying stones to pave streets and railroads.

Lieutenant Colonel Robert Maginnis (U.S. Army, retired) shares Solzhenitsyn's concern and compassion for women put into physically demanding, dirty, and traditionally male-roles, specifically combat. Maginnis (2013:193-194) poignantly asks what kind of country are we to subject our women to combat? He also notes that the policy of integration of women in combat is unprecedented in human history, should be analyzed by knowledgeable people, and policy should prioritize military effectiveness (Maginnis 2013:187).

The biology of the sexes should have been a primary consideration in the decision to integrate women into all MOS, including combat MOS. This was apparently not the case, as indicated in Secretary of Defense Ash Carter's justifying the policy by claiming that we need to take full advantage of all individuals, including women, who can meet military standards (Carter 2015). Allowing women into all military MOS if they meet standards may seem logical. However the focus on standards disregards the fact that sex and reproduction are among the most basic factors in human biology. If the basic biology of the sexes has not been thoroughly considered then the risks of sexual integration have not been adequately assessed.

I think that Secretary Carter's claim that we have to take advantage of all individuals who meet standards is contrived. If he meant what he said about using every person who can meet standards, then he should have also included men older than the current military age limits if they meet standards. Of course, the traditional policies on age limits and exclusion of women from combat MOS were instituted for a reason. I think the historical pattern with restrictions on age and sex in the military developed in part because of the superior physical abilities and aggression of young men, and also because group dynamics and cohesion are critical to military structure. Junger (2016) provides good insights on male group dynamics and war.

Most women and older men are not as physically capable or aggressive as younger men, and women or older men would significantly change the dynamics and cohesion of groups composed primarily of young men. The military is structured the way it is, a strict hierarchy of men with explicit display of rank (analogous to dominance displays in animals) and group (i.e., the service branch and unit) for a reason. It has developed over time and it works. It is relevant that even in cases where women have been pressed into combat, this system is maintained. For example, most women in the Soviet Army in World War II were in all-female units (Fukuyama 1998, Browne 2007:279). See also Balke's (2017) discussion of Female Engagement Teams (FET) currently in use in the U.S. military.

General John Kelly (U.S. Marine Corps) was asked about the policy to integrate the sexes with regard to Marine Corps reports that found mixed gender units were less lethal, slower, and more prone to injuries than all-male units. He responded in part (Kelly 2016):

> "...I think every decision has to be looked at (with) only one filter...does it make us more lethal on the battlefield?...will it result in less casualties on our side?... the only science I know on this ... was the study that the Marine Corps contracted the University of Pittsburgh ... because of the nature of infantry combat, infantry training, and all of the rest, there's a higher percentage of young women in the scientific study that get hurt, and some of them get hurt forever."

The science to which General Kelly referred was dismissed by Secretary Carter who ordered the full integration of the sexes in the military anyway. This science includes empirical studies that compared performance in combat-related tasks of male-only and male-female integrated units, and indicate that men are more qualified for combat than women (U.S. Marine Corps 2015, Balke 2017). Equally as important, other

81

studies show that pregnancy, menses, and sexual attraction between men and women can affect the dynamics and cohesiveness of military units, potentially resulting in reduced effectiveness and increased casualties (Mitchell 1989, 1998, van Creveld 1991, 2001, Gutmann 2000, Browne 2007, Maginnis 2013). These studies are relevant to the military's job of preventing wars if possible, and winning them with minimum casualties if not. However, those making the policy (which the military must now implement) dismissed them. Policymakers should have been aware of all of the relevant science for a policy that is unprecedented in history and has life and death consequences for many people and the nation. An objective review of this information does not lead to a conclusion that integrating the sexes in combat MOS will make the military more effective. The policy makers should have been aware this.

The point of this discussion is that men and women are biologically different, and this should have been acknowledged before making the decision to integrate the sexes. Now, four years later, the special needs of women in the military regarding medical issues and equipment (e.g., ill-fitting body armor) are becoming quite apparent (e.g., O'Connor and Bergengruen. 2019, Seck 2018). These problems and others were well-documented prior to the decision to integrate the sexes (Mitchell 1989, 1998, Van Creveld 1991, 2001, Gutmann 2000, Browne 2007, Maginnis 2013).

It is also worth considering the effect of integration of the sexes on men. Men have traditionally been the protectors and warriors of society and they want recognition from women as such (as indicated in the quote by Lawrence above). Being strong, brave, and a warrior gives men pride and they want women to recognize and appreciate it. Do women serving in the same roles as men diminish men's pride and motivation? This question is worthy of serious consideration.

I wrote an essay on this subject in *Small Wars Journal* (Cronin 2018) to stimulate discussion of the basic biology of the sexes as it relates to military integration.

I pointed out in the essay that the literature on women in the military (Mitchell 1989, 1998, Gutmann 2000, van Creveld 2001, Browne 2007, Maginnis 2013) shows that this policy began with efforts by feminist lobbyists to get equal rights for women. However, denying or minimizing the differences between the sexes became part of the narrative as described in feminist literature.

For example, Itzin (1992:13) claimed that male characteristics of dominance, aggression, and masculinity "...are not biologically determined." Instead, she claimed that masculine traits are conditioned and can be deconditioned, and that men can change.

Imagine this claim referring to male cattle (bulls), male chimpanzees, or any other male animal that fights for dominance and mates, and its absurdity is apparent (at least to a biologist). Men and women are different and the differences *are* biologically determined by genetics. This should be explicitly recognized in discussions about integration of the sexes into combat MOS and the military in general.

Marine Corps Captain Lauren Serrano (2014) expressed this well in the *Marine Corps Gazette*, in which she makes the case that women do not belong in the infantry. She noted that women are different than men, and not just physically different. She also recognized that this needs to be acknowledged in the discussion of integration of the sexes.

Assessment of sexual integration of the military has shown an effect on readiness and morale, and problems with sexual harassment of women by men (Mitchell 1989, 1998, Gutmann 2000, van Creveld 2001, Browne 2007, Maginnis 2013). These assessments have revealed a "great paradox" in which it is claimed that men and women are the same and they can be integrated; but there are extensive rules and exceptions to provide special treatment of women and keep them segregated (Gutmann 2000:209).

Perhaps the most important factor in this discussion is the historical pattern that men fight wars, as described by John Keegan (1993:75-76). He explains, with the authority of an eminent historian, that historically women did not fight in war. Men fought men in war. Women helped men in war, but have not been combatants. The exceptions in which women have been pressed into combat are rare and anomalous, as explained in detailed assessments (Mitchell 1989, 1998, Gutmann 2000, van Creveld 2001, Browne 2007, Maginnis 2013).

Consideration of more ancient history, our evolutionary history, is also relevant as I discussed in my *Small Wars Journal* article:

"Science also suggests that the historical pattern of men fighting other men in war is rooted in our evolutionary history...

Despite our complex civilization, humans retain the genetic legacy of our primitive ancestors that were subject to natural selection (commonly known as survival of the fittest) in hunting and warfare...

Humans, before modern society with its abundant food and medicine, had high infant mortality and short life spans. There was strong selection for traits (biological and cultural) that contributed to the production and survival of children to reproductive age. In our ancestral hunter-gatherer societies, men were the primary hunters and warriors and women gathered food and bore, nursed, and raised children. Such sexual division of labor was efficient and conferred fitness to the group. These conditions might or might not apply to modern society, but they are part of our biological and cultural heritage."

Human history shows that women participated in war in supporting roles and not as combatants. The risks associated with the fundamental change of integration of the sexes should be assessed considering science that indicates different physical attributes and roles of men and women in society developed in nature and enhanced fitness of individuals and populations. Changing the historical pattern by allowing women to be combatants in war entails risks to the basic structure of human populations, as well as military effectiveness. This indicates that the risks associated with full integration of the sexes in the military have not been adequately assessed.

Solutions

Potential solutions to the issues of integration of the sexes in the military:

1. Explicitly recognize that men and women are biologically different and reject feminist claims to the contrary. Ensure that all personnel in the military are educated in this regard.

2. Explicitly recognize that men, and not women, have historically been the combatants in war for legitimate, biologically-based reasons. Ensure that all personnel in the military are educated in this regard.

3. Maintain the current policy of full integration of women, but acknowledge points 1 and 2 above.

4. Revert to the policy excluding women from combat MOS.

5. If the policy of integration of women into combat MOS is maintained, assign women to all-women segregated units, such as the existing Female Engagement Teams (FET).

6. Allow military experts to determine the policy of integration of the sexes in the military.

85

Chapter 8 God and Science
Bullet Summary

— Mankind struggles with the question of whether God exists or does not exist.

— The existence of God is considered a scientific question or a non-scientific spiritual question by different people.

— Some people think God does not exist based on lack of scientific evidence.

— Some people think God does exist based on science.

— We do not know with scientific certainty if God exists or does not exist.

— Science provides strong evidence that humans evolved from ancestors with less mental capacity than modern humans.

— Humans at present may not have the mental capacity to comprehend all aspects of nature, the universe, and God.

— Human evolution included retention of juvenile characteristics (known as paedomorphosis and neoteny) that resulted in larger brains and greater mental capacity in modern humans than in our evolutionary ancestors.

— Future evolution could result in increased mental capacity that allows better comprehension of the universe and God.

— Our current understanding of science and human evolution does not support literal interpretation of creation happening in seven days as described in Genesis.

— Evolutionary analogies to Genesis provide insights regarding creation and the future evolution of mankind's mental capacity to comprehend God.

— Christ telling us to be like children and identifying Himself as the Son of Man is consistent with the evolution of increased mental capacity through retention of juvenile characteristics that can ultimately result in human capacity to understand God.

Chapter 8

GOD AND SCIENCE

In this chapter I discuss aspects of science, religion, evolution, and the question of God's existence. This topic is relevant to the theme of this book that there has been government control of this issue, but not in the U.S. where the First Amendment protects our religious beliefs. However, as with environmentalism and feminism, there is a connection of belief or non-belief in God and socialism/communism. The Soviet Union was a state based on atheistic principles with government control of religion that resulted in a disaster for human rights and freedom (Glynn 1997:161).

Scholars have debated the existence of God throughout history. The Resurrected Jesus said to Thomas:

> "Blessed are those who have not seen and yet believe.".
> (John 20:29).

Belief versus non-belief in God has been a constant question for humanity. A person who believes in God might say: I have not seen, but I believe. An atheist might say I have not seen, so I do not believe. However, believing and not believing in God are both beliefs. As noted by physicist Marcelo Gleiser (2019) "atheism expresses belief in nonbelief".

Some people think that science supports non-belief. A recent assessment by Richard Dawkins (2006) in his book *The God Delusion* includes a chapter entitled: "Why there almost certainly is no God". Dawkins' analysis is an endorsement of atheism, with an extensive and informed presentation of science, history, and religion. He treats the existence or non-existence of God as a scientific question and demonstrates that science does not support the existence of God. He acknowledges in the chapter title that we are not 100% certain there is no God, as is appropriate for a scientist. Of course, we are also not 100% certain there is God. Other authors have also discussed science and the existence of God (e.g., Davies 1992, Behe 1996, Glynn 1997, Gleiser 2019). Another issue involving science is the debate over creationism and evolution (see Johnson 1993, Futuyma 1995, Roughgarden 2006).

Regarding the premise that God's existence is a scientific question, consider that one can believe that the sun will rise tomorrow (i.e., the Earth will rotate and the sun will remain illuminated). Science based on innumerable observations of the sun rising each morning supports this belief, but we don't know with 100% certainty. One can also believe there is no God based on lack of scientific evidence but doesn't know with 100% certainty.

I think there is a fundamental difference between belief in the sun rising (it is likely based on science) and belief in the existence of God (it is arguably unlikely based on science). The difference isn't the argument that the sun rising is a scientific question, and that God's existence is not a scientific question but a spiritual one. Nor is it that the sun rising tomorrow is a prediction about the future, and God's existence or non-existence is in the present. The difference is that our senses and mental capacities can perceive and comprehend the sun rising, but perhaps our senses and mental capacities *are not able* to perceive and comprehend God. I think Albert Einstein expressed this idea:

"I don't try to imagine a personal God; it suffices to stand in awe at the structure of the world, insofar as it allows our inadequate senses to appreciate it."

In other words, we simply may not be able to comprehend things (that we call science) that reveal whether or not God exists. Of course there is variation in people's ability to comprehend different things (e.g., mathematical or musical ability), but there is a limit to everyone's mental capabilities, as Einstein recognized. This is perhaps an argument for agnosticism (i.e., the view that the reality of God is unknowable).

However, consider that extant (i.e., currently existing) humans evolved from ancestors with different, and presumably less, mental capability. Our ancestors likely could not comprehend some things that we can. I think it follows that extant humans have limits to our comprehension, and we might have mental capacity less than that required to comprehend everything about the universe, including God. Maybe we *can't* understand God in scientific terms at this point of our evolution. Perhaps the future mental capacity of our evolutionary descendants will be greater than ours. I do not know if humans are still evolving mental capacity but presumably it is likely to occur over long periods of time, as it has in the past. So maybe the ultimate stage of human evolution is acquiring the mental capacity to comprehend the currently incomprehensible existence and nature of God. I think this view justifies belief in God based on faith and science.

Eden and Evolution

Our current understanding of science and human evolution does not support literal interpretation of creation happening in seven days as described in Genesis. The Biblical account of creation is considered as "myth... the primeval history of humans...the process of creation and ideas about the earliest human ancestors as the ancients understood them." (The Catholic Study Bible 2006, Reading Guide page100).

Carl Sagan (1977) in his book *The Dragons of Eden* describes the Genesis account of the Garden of Eden as a metaphor for evolution. He notes that the increase in the size of the female human pelvis evolved to accommodate birth of babies with large heads, large brains, and superior intelligence. Further increase of the size of the pelvis would likely impair women's ability to walk and run.

God says to Eve as punishment for eating fruit from the tree of the knowledge of good and evil:

"I will intensify your labor pains; you will bear children with painful effort." (Genesis 3:16).

Sagan describes how the pain of childbirth endured by women is due to the rapid evolution of increased size of the human skull and brain. The large head passing through the birth canal is painful, and the size of the pelvis increased to accommodate this as much as possible. The neocortex of the brain developed with the increase in brain size, and the neocortex is likely where abstract and moral judgements occur. Eden represents the beginning of mankind with what we now consider human consciousness accompanied with God's bestowal to mankind of a soul (the existence of a human soul is another issue of belief or non-belief).

I offer the following ideas and insights about the events in the Garden of Eden and Christ in the context of human evolution. Some of these ideas are similar to Sagan's and may be similar to those of others of which I am not aware. My insights follow the idea I stated above: that human mental capacity to comprehend God is currently limited, but can potentially increase through evolution.

The story of the Garden of Eden in Genesis is the story of the origin of the human race (The Catholic Study Bible 2006, Introduction to Genesis page 6), and can be seen as a metaphor for the evolution of large brains, intelligence, self-consciousness, and awareness of death and mortality. These characteristics separated man from other animals and our human ancestors, and were accompanied by an increased mental capacity that included the conception of God.

Large brains developed with larger heads, and the female pelvis widened as much as possible without severely impairing mobility, as Sagan describes. The mechanism of this evolution is often considered heterochrony (i.e., changes in rates of development of different parts of the body). This may be achieved with changes in timing of gene regulation, activation, and inactivation. Paedomorphosis is a type of heterochrony in which juvenile traits are maintained in the reproductively mature adult. An example is salamanders that retain gills from the juvenile tadpole stage into the adult stage. Paedomorphosis can occur through retardation of somatic growth (i.e., non-reproductive tissues which is most of the body, called neoteny); or acceleration of reproductive growth (called progenesis). The science of heterochrony, development, and evolution is complex with a large literature (see Gould 1977, Bergstrom and Dugatkin 2016, Futyuma and Kirkpatrick 2017). Francis (2015) provides a lucid explanation of neoteny in humans and animals.

Modern humans have been called "neotenic apes" considering several characters that were likely present in juveniles of our ancestors and retained in adult modern humans (Gould 1977). Apes are our closest living evolutionary relatives, but it is important to remember that extant apes are not our ancestors. They evolved concurrently with, but independently of, humans from an ape-like common ancestor that existed about six to seven million years ago (Young et al. 2015).

Neoteny in humans has been attributed to several characteristics including longer childhood, greater flexibility of behavior, larger heads and brains, less developed bodies at birth, and less body and facial hair compared with human ancestors and apes. If humans are neotenic compared to our ancestors, in a sense we can be considered "childlike".

One of the most childlike characters relevant to the Eden story is curiosity (called exploratory by psychologists). Children are generally more curious and have greater plasticity of behavior and ability to learn than adults. This has been called behavioral neoteny (Lorenz 1971, Gould 1977:402). It was Adam and Eve's curiosity, wanting to know as much as God, that led to their fall. Admittedly, God wasn't pleased with Adam and Eve's *acting* on their curiosity, but He bestowed this trait on them, so presumably He wanted them to be curious.

Humans have also been called "naked apes" (Morris 1967) because of our lack of hair compared to apes. This is considered another juvenile trait in humans compared to our ancestors. Recall that Adam and Eve became aware and ashamed of their nakedness after eating the forbidden fruit providing another metaphor of creation and evolution.

It is therefore relevant that Jesus wants us to be like children:

> "Truly I tell you, unless you turn and become like children, you will never enter the kingdom of heaven. Therefore, whoever humbles himself like this child-this one is the greatest in the kingdom of heaven." (Matthew 18:3-4).

The point is that Eden represents the beginning of conscious mankind, some of the traits in the first humans described in Genesis can be analogized to current evolutionary thought about neoteny and juvenile characteristics, and Jesus tells us to be childlike.

There is also relevance of other words of Christ in this regard. Jesus was the self-identified "Son of Man" (see Matthew 12:8 in the Literature Cited for additional verses referencing the Son of Man). Jesus as the Son of God is understandable, but the meaning of Son of Man is somewhat mysterious. Craig Blomberg, in an interview with Lee Strobel (1998:36), describes how the Son of Man is foretold in the Old Testament (Daniel 7:13) and that it indicates Jesus' divinity. My idea is that Jesus, as the Son of Man, is the example of the evolutionary *descendant* (a son is a descendant) of the Old Testament man, with a new creed of loving one's neighbors and enemies.

I think Jesus' message is that salvation depends on following His example of selflessness and love of neighbor and enemy alike by being childlike. It is a message of what mankind needs to become to avoid death and damnation. In evolutionary terms this instructs us how to attain fitness (i.e., survival and reproductive success) and avoid extinction. Immortality can be seen as successfully producing descendants and avoiding extinction (although I don't imply this excludes immortality in an afterlife in Heaven, another issue of belief or non-belief).

Children are humble and loving, and Jesus' message to be like children is consistent with human evolution of juvenile characteristics (i.e., paedomorphosis and neoteny). As the Son of Man, Jesus is also showing us that we must emulate Him and be childlike descendants of ancestral men.

Just as our evolution to be more childlike, by retention of juvenile traits, resulted in an increase in mental capability, future evolution with Jesus as a model can result in a change, presumably an increase, of our mental capacity. As the Son of Man, He may be indicating that our evolutionary descendants will have increased mental capacity (gained through the retention of juvenile traits) and will ultimately be able to comprehend the mysteries and truth about the existence of God.

Chapter 9 Socialist, Communist, Marxist Science, and Lysenkoism
Bullet Summary

— The Soviet Union controlled science to fit the socialist/communist government's agenda during the infamous period of Lysenkoism.

— In the case of Lysenko it was agricultural science.

— Lysenkoism endorsed inheritance of acquired characteristics and rejected Mendelian genetics to be consistent with socialism.

— Lysenko's incorrect science resulted in large failures of Soviet agriculture which resulted in the deaths of millions of people.

— Dissent from Lysenko's theories was outlawed in the Soviet Union in 1948, and dissenters were dismissed from jobs, imprisoned, or sentenced to death as enemies of the state.

— This included an accomplished botanist, Nikolai Vavilov, who died in prison.

— Lysenkoism has been described as a warning of the dangers of bureaucratic and ideological distortions of science.

— Lysenkoism was made possible by a totalitarian socialist/communist government.

— Some Americans are now endorsing socialism in the U.S.

— Modern science in the U.S. is not descending into Lysenkoism, modern science is rigorous, but science can be distorted by political agendas.

— Authors have warned about the threat of Lysenkoism reemerging today for environmental issues such as global warming.

— Americans should be aware of history, and oppose government control of science and stifling dissent to achieve policy agendas.

Chapter 9

SOCIALIST, COMMUNIST, MARXIST SCIENCE, AND LYSENKOISM

A leksandr Solzhenitsyn and I never met, but our eyes did. It was as I was boarding a plane. Probably Alaska Airlines, Seattle to Anchorage in the 1990's. It was around the time Solzhenitsyn was returning to Russia in 1994 after his exile in Vermont, so he was probably on his return journey. As I remember, he was sitting in a first class seat, port side of the plane. I looked down as he looked up. Piercing, deep eyes, blue as I recall, with knowledge, sadness, and, the best word I can find, friendship. He was writing, in Russian I suppose. Clear neat script. Like my Dad's. I was taken aback, by his eyes and the realization of who this was. I may have mumbled "Hi" or some other greeting and continued to my seat. This man had been through World War II, the Soviet Gulag and persecution, stood up to the communist Soviet leaders, and survived physically, mentally, and motivationally to write about it and receive the Nobel Prize for literature. He had guts.

Although my encounter with Solzhenitsyn was brief, it made me better appreciate his insights as expressed in his writings. In his acceptance speech for the Nobel Prize (Solzhenitsyn 1972), he lamented that young people at that time were enthusiastic about repeating the history that led to communism with its total government control; and the reluctance of more experienced people to refute them. He was warning against ignoring history, or being condemned to repeat it.

95

The Soviets controlled science to fit the communist government's agenda during the infamous period of Lysenkoism. The eugenics movement of the early 1900's including Hitler's racist policies is another example of government controlling science to advance a political agenda (see Crichton 2004). In the case of Lysenko it was agricultural science. Lewontin and Levins (1976) discuss "Marxist science" and how Lysenkoism was complex, but it was a failure and set Soviet genetics back a generation. They note that Lysenkoism is an example of the danger of distortions of science by bureaucracy and ideology. Other authors have described Lysenko in great detail (Medvedev 1969, Joravsky 1970, Soyfer 1994, Birstein 2001, Roll-Hansen 2005).

Trofim Denisovich Lysenko was from a Ukrainian peasant family, and had Stalin's support and became director of the Institute of Genetics in the USSR Academy of Sciences. He had great influence on Soviet agriculture from the 1930s to the 1960s. Lysenko's scientific ideas included inheritance of acquired characteristics, a concept known as Lamarckism, named after the French naturalist Jean-Baptiste Lamarck of the 1700-1800s. Lamarck contributed good work to biology at a time when genetics and heredity were not understood. He held the belief, common in his day, that a trait acquired during an organism's lifetime could be inherited by its descendants. Lamarck and his ideas predated both Charles Darwin and Gregor Mendel, whose findings replaced Lamarck's ideas and set the groundwork for modern genetics.

Lysenko's views on the inheritance of acquired characteristics were consistent with socialist/communist doctrine that heredity and genetics has a limited role in determining human characteristics. Instead, they emphasized the environmental influences on many traits, and rejected, or misunderstood, Mendelian genetics and Darwinian natural selection. The relationship of Lysenkosim, Mendelian genetics, and Darwinism is complex and thorough assessments should be consulted for a full understanding (Medvedev 1969, Joravsky 1970, Lewontin and Levins

1976, Soyfer 1994, Birstein 2001, Roll-Hansen 2005). Some communists thought that people could be changed under socialism/communism and these changes would be inherited in subsequent generations resulting in new selfless, socialist men (Ferrara 2013). The idea that the environment and not heredity, is the primary determinant of traits was also taken to absurd extremes by feminist claims that male traits could be conditioned and were not biologically determined, as noted in Chapter 7. The emphasis on environmental conditioning is shared by feminism and communism.

Lysenko's rejection of legitimate science (e.g., Mendelian genetics) resulted in large failures of Soviet agriculture and famines that killed millions of people. Communist China also tried Lysenkoist agriculture (Schneider 1986).

Dissent from Lysenko's theories was outlawed in the Soviet Union in 1948. Soviet scientists who would not accept Lysenko's science were dismissed from their jobs, many were put in prison, and some were sentenced to death as enemies of the state. This included the accomplished botanist, Nikolai Vavilov, who died in prison in 1943. Following Stalin's death and the rejection of Lysenko's science in the Soviet Union, Vavilov's reputation was restored as a national hero in the 1960s.

Modern science in the U.S. is not descending into Lysenkoism. As noted in Chapter 1, modern science is rigorous and performance-driven. However, the distinction between science and government policy can be distorted by political agendas. In this regard, it is important to acknowledge that environmentalism has become a primary tool of socialists/communists trying to force more government control over the economy and land owners (see Arnold 2007, Horner 2007, Andre 2011). Others have warned about the threat of Lysenkoism reemerging today (Andrews 2006), particularly for environmental issues and global warming (Crichton 2004, Ferrara 2013). These authors warn about the dangers of using science to advance political agendas, and similarities of today's political climate and Lysenkoism. Like Crichton, Ferrara, and Andrews, I present the case of Lysenkoism in this book to warn against the potential for the government to control science for policy agendas and to stifle dissent. The government control of science with the Endangered Species Act and other issues that I discuss is not as severe as Lysenkoism which was based on incorrect science and banning dissent. My experience is with the selective use of science by the government, and I think it is important to remind Americans of the Soviet experience so as not to repeat it.

Lysenkoism was made possible by a totalitarian socialist/communist government. Recall that the Soviet Union's name was the Union of Soviet *Socialist* Republics (USSR) and it was known as a socialist and a communist country. This is relevant because even though "The Cold War is Over" (Hyland 1990) and the Soviet Union dissolved, some Americans are now calling for socialism (see Kesler 2018), as the youth lamented by Solzhenitsyn did in in 1972. I suppose each generation has a component attracted to socialism/communism with its promise of full equality, without recognizing its inevitable descent into government control and loss of liberty.

I believe that Americans supporting socialism and calling for government control on many issues is cause for concern, and even alarm. The episode of Lysenkoism in the Soviet Union should be a warning of the potential terrible results of such policies. I hope acknowledgement of the misuse of science in history will prevent socialist/communist control of science and infringement on American's liberty and well-being in the future. It is our responsibility to oppose such government control of science.

LITERATURE CITED

Note: The websites in citations were accessed in June 2019.

Anderson, T.L. Editor. 2000. Political Environmentalism: Going Behind the Green Curtain. Hoover Institution Press, Stanford University, Stanford CA.

Andre, J. 2011. U.S.A. vs E.S.A. The Politically Incorrect Side of the Endangered Species Act of 1973. Published by John Andre, Hamilton, Montana. ISBN-13: 978-1466431393

Andrews, J. 2006. Lessons from Lysenko. Phytopathology News 40 (6):66.

Arnold, R. 2007. Freezing in the Dark: Money, Politics and the Vast Left Wing Conspiracy. Merril Press. Bellevue, Washington.

Avise, J.C. 2000. Phylogeography, the History and Formation of Species. Harvard University Press, Cambridge, Massachusetts.

Avise, J.C. and R.M. Ball Jr. 1990. Principles of genealogical concordance in species concepts and biological taxonomy. Oxford Survey of Evolutionary Biology 7:45–67.

Baker, R.J. and R.D. Bradley. 2006. Speciation in mammals and the genetic species concept. Journal of Mammalogy 87:643-662.

Balke, L.A. 2017. Gender-Integration in the United States Marine Corps. A report submitted in partial fulfillment of the requirements for the degree of Master of Military Studies. United States Marine Corps Command and Staff College, Marine Corps University, Quantico, Virginia.

Ballard, W.B., M.A. Cronin, and H.A. Whitlaw. 2000. Interactions between caribou and oil fields: the Central Arctic caribou case history. Pages 85-104 In: The Natural History of an Arctic Oil Field:

Development and the Biota. J.C. Truett and S.R. Johnson Editors. Academic Press, San Diego.

Barrowclough, G.F., R.J. Gutierrez, and J.G. Groth. 1999. Phylogeography of spotted owl populations based on mitochondrial DNA sequences: gene flow, genetic structure, and a novel biogeographic pattern. Evolution 53:919-931.

Barrowclough, G.F., J.G. Groth, L.A. Mertz, and R.J. Gutierrez. 2005. Genetic structure, introgression, and a narrow hybrid zone between northern and California spotted owls. Molecular Ecology 14:1109-1120.

Behe, M.J. 1996. Darwin's Black Box: The Biochemical Challenge to Evolution. The Free Press, a Division of Simon & Schuster, New York.

Benjamin, D.K. 2004. Incentives and Conservation: The Next Generation of Conservationists. The Property and Environment Research Center (PERC), Bozeman, Montana.

Bergstrom, C.T. and L.A. Dugatkin. 2016. Evolution 2nd edition. W.W. Norton & Company, New York.

Birstein, V.J. 2001. The Perversion of Knowledge: The True Story of Soviet Science. Westview Press, a Member of the Perseus Books Group, Boulder, Colorado.

Bosworth, D. and H. Brown. 2007. After the timber wars: Community-based stewardship. Journal of Forestry 105(5):271-273.

Browne, K. 2007. Co-Ed Combat: The New Evidence that Women Shouldn't Fight the Nation's Wars. Sentinel, The Penguin Group, New York.

Bureau of Land Management. 2002. Environmental Impact Statement for the Renewal of the Federal Grant for the Trans-Alaska Pipeline System Right-of-Way. Bureau of Land Management, Department of the Interior, Washington, D.C. This EIS was prepared by the U.S. Department of Energy Argonne National Laboratory.

Bureau of Land Management. 2018. Coastal Plain Oil and Gas Leasing Program Draft Environmental Impact Statement Bureau of Land Management, Department of the Interior. December 2018. Washington, D.C.

Carter, A.B. 2015. Department of Defense Press Briefing by Secretary of Defense Ash Carter December 3, 2015. https://dod.defense. gov/News/Transcripts/Transcript-View/Article/632578/depart-ment-of-defense-press-briefing-by-secretary-carter-in-the-penta-gon-briefi/

Cavalli-Sforza, L. L., P. Menozzi, and A. Piazza. 1994. The History and Geography of Human Genes. Princeton University Press, Princeton, New Jersey.

The Catholic Study Bible. 2006. D. Senior and J.J. Collins Editors. Oxford University Press, New York.

Chambers, S.M., S.R. Fain, B. Fazio, and M. Amaral. 2012. An account of the taxonomy of North American wolves from morphological and genetic analyses. North American Fauna 77:1-67.

Chase, A. 1987. Playing God in Yellowstone. Harcourt Brace Jovanovich, Orlando Florida.

Chase, A. 1995. In a Dark Wood: The Fight Over Forests and the Rising Tyranny of Ecology. Houghton Mifflin Company, Boston and New York.

Clausewitz, C. von. 1832. On War. Originally *Vom Kriege* published in 3 volumes, Berlin. Republished in 1976. Translated by M. Howard and P. Paret. Princeton University Press, Princeton, New Jersey.

Coffman, M. 1994. Saviors of the Earth? The Politics and Religion of the Environmental Movement. Northfield Publishing, Chicago, Illinois.

Cooper, C.R. 2018. Court Opinion on a petition to list bison in Yellowstone National Park under the Endangered Species Act. U.S. District Court for the District of Columbia. Buffalo Field Campaign et al., Plaintiffs, v. Ryan Zinke et al., Defendants. Case 1:16-cv-01909-CRC, Document 27 Filed 01/31/18.

Cracraft, J. 1989. Speciation and its ontology. Pages 28-59 In: Speciation and its Consequences. D. Otte and J. Endler Editors. Sinauer Associates, Sunderland, Massachusetts.

Crichton, M. 2004. State of Fear: Appendix I. Harper Collins, New York.

Crockford, S.J. 2019. The Polar Bear Catastrophe that Never Happened. The Global Warming Policy Foundation, London, United

Kingdom. See also the Polar Bear Science website: https://polar-bearscience.com/

Cronin, M.A. 1996. Advocacy inappropriate for public agencies. Wildlife Society Bulletin 24:2.

Cronin, M.A. 1997. Systematics, Taxonomy, and the Endangered Species Act: The Example of the California Gnatcatcher (*Polioptila californica*). Wildlife Society Bulletin 25:661-666.

Cronin, M.A. 2001. Did Federal Agency Set Up Norton? Voice of the Times, Anchorage Daily News, November 15, 2001.

Cronin, M.A. 2006. A Proposal to eliminate redundant terminology for intra-species groups. Wildlife Society Bulletin 34:237-241.

Cronin, M.A. 2007a. EIEIO: Evolutionary Interesting and Ecologically Important Organisms. Range Magazine Spring 2007:24-25.

Cronin, M.A. 2007b. The Preble's meadow jumping mouse: subjective subspecies, advocacy and management. Animal Conservation 10:159-161.

Cronin, M.A. 2013. Federal oil for the U.S. military: A solution to the impact of sequestration on the military, enhanced national security, and economic growth. Unpublished essay, University of Alaska Fairbanks, School of Natural Resources and Agricultural Sciences, Agricultural and Forestry Experiment Station, Palmer, Alaska, September 11, 2013.

Cronin, M.A. 2015. The Greater Sage-Grouse Story: Do we have it right? Rangelands 37:200-204.

Cronin, M.A. 2017. Written Testimony of Matthew A. Cronin for the U.S. Senate Energy and Natural Resources Committee, on The Potential for Oil and Gas Exploration and Development in the Non-Wilderness Portion of the Arctic National Wildlife Refuge, Known as the "1002 Area" or Coastal Plain, to Raise Sufficient Revenue Pursuant to the Senate Reconciliation Instructions included in H. Con. Res. 71. November 2, 2017. The Hearing can be viewed, and written comments downloaded, at either of these two websites: https://www.energy.senate.gov/public/index.cfm/hearings-and-business-meetings?ID=97CD66D2-BD9D-49AD-87EB-DE4C68A888DF;

https://www.energy.senate.gov/public/index.cfm/2017/11/full-committee-hearing

Cronin, M.A. 2018. Integration of the sexes in the military: Biological considerations. Small Wars Journal 15 June 2018. http://smallwarsjournal.com/jrnl/art/integration-sexes-military-biological-considerations.

Cronin, M.A. 2019a. The Endangered Species Act and Science. Western Livestock Journal February 5, 2019. https://www.wlj.net/opinion/guest_opinion/resource-science-the-endangered-species-act-and-science/article_89f251f8-25a6-11e9-b51f-97680dea780d.html

Cronin, M.A. 2019b. Bison and the Endangered Species Act. Western Livestock Journal, April 22, 2019. https://www.wlj.net/opinion/guest_opinion/resource-science-bison-and-the-endangered-species-act/article_a75869da-6525-11e9-95fa-47d29f1f9fb3.html

Cronin, M.A. 2019c. Grizzly bears, polar bears, and the ESA. Western Livestock Journal, March 18, 2019. https://www.wlj.net/opinion/guest_opinion/resource-science-grizzly-bears-polar-bears-and-the-esa/article_be98912a-5561-11e9-8c78-b72662406322.html

Cronin, M.A. 2019d. Wolf species, subspecies, populations, and the ESA. Western Livestock Journal, February 18, 2019. https://www.wlj.net/opinion/guest_opinion/resource-science-wolf-species-subspecies-populations-and-the-esa/article_b25ee890-33a2-11e9-8bb4-e77046434f29.html.

Cronin, M.A. and L.D. Mech. 2009. Problems with the claim of ecotype and taxon status of the wolf in the Great Lakes region. Molecular Ecology 18:4991-4993.

Cronin, M.A. and M.D. MacNeil. 2012. Genetic relationships of extant North American brown bears (*Ursus arctos*) and polar bears (*U. maritimus*). Journal of Heredity 103:873-881.

Cronin, T.M. and M.A. Cronin. 2015. Biological Response to Climate Change in the Arctic Ocean: the View from the Past. Arktos: The Journal of Arctic Geosciences. Review article. DOI 10.1007/s41063-015-0019-3

Cronin, M.A., S.C. Amstrup, G.W. Garner, and E.R. Vyse. 1991. Interspecific and intraspecific mitochondrial DNA variation in North American bears (*Ursus*). Canadian Journal of Zoology 69:2985-2992.

Cronin, M.A., B.J. Pierson, S.R. Johnson, L.E. Noel, and W.B. Ballard. 1997. Caribou population density in the Prudhoe Bay region of Alaska. The Journal of Wildlife Research 2:59-68.

Cronin, M.A., S.C. Amstrup, G.M. Durner, L.E. Noel, and W.B. Ballard. 1998a. Caribou distribution during the post-calving period relative to oil field infrastructure in the Prudhoe Bay oil field, Alaska. Arctic 51:85-93.

Cronin, M.A., W.B. Ballard, J.D. Bryan, B.J. Pierson, and J.D. McKendrick. 1998b. Northern Alaska oil fields and caribou: A commentary. Biological Conservation 83:195-208.

Cronin, M.A., H.A. Whitlaw, and W.B. Ballard. 2000. Northern Alaska oil fields and caribou. Wildlife Society Bulletin 28:919-922.

Cronin, M.A., H.A. Whitlaw, and W.B. Ballard. 2001. Addendum. Northern Alaska oil fields and caribou. Wildlife Society Bulletin 29:764.

Cronin, M.A., M.D. MacNeil, and J.C. Patton. 2005. Variation in mitochondrial DNA and microsatellite DNA in caribou (*Rangifer tarandus*) in North America. Journal of Mammalogy 86:495-505.

Cronin, M.A., M.D. MacNeil, N. Vu, V. Leesburg, H. Blackburn, and J. Derr. 2013a. Genetic variation and differentiation of extant bison subspecies and comparison with cattle breeds and subspecies. Journal of Heredity 104:500-509.

Cronin, M.A., M.M. McDonough, H.M. Huynh, and R.J. Baker. 2013b. Genetic relationships of North American bears (*Ursus*) inferred from amplified fragment length polymorphisms and mitochondrial DNA sequences. Canadian Journal of Zoology 91:626-634.

Cronin, M.A., G. Rincon, R.W. Meredith, M.D. MacNeil, A. Islas-Trejo, A. Canovas, and J.F. Medrano. 2014. Molecular phylogeny and SNP variation of polar bears (*Ursus maritimus*), brown bears (*U. arctos*) and black bears (*U. americanus*) derived from genome sequences. Journal of Heredity 105:312-323.

Cronin, M.A., A. Cánovas, A. Islas-Trejo, D.L. Bannasch, A.M. Oberbauer, and J.F. Medrano. 2015a. Single nucleotide polymorphism (SNP) variation of wolves (*Canis lupus*) in Southeast Alaska and comparison with wolves, dogs, and coyotes in North America. Journal of Heredity 106:26-36.

Cronin, M.A., A. Cánovas, A. Islas-Trejo, D.L. Bannasch, A.M. Oberbauer, and J.F. Medrano. 2015b. Wolf Subspecies: Reply to Weckworth et al. and Fredrickson et al. Journal of Heredity 106:417-419.

Darwin, C. 1871. The Descent of Man and Selection in Relation to Sex. Murray, London. Republished in 1981, Princeton University Press, Princeton, New Jersey. The quote on women and war is in Volume II Sexual Selection page 323 and is evidently from Horace, Satires, I.iii.107–110.

Davies, P. 1992. The Mind of God: The Scientific Basis for a Rational World. Touchstone, Simon & Schuster, New York.

Dawkins, R. 2004. The Ancestor's Tale. Houghton Mifflin Company, New York.

Dawkins, R. 2006. The God Delusion. Mariner Books, Houghton Mifflin Harcourt, Boston and New York.

Derocher, A.E. 2012. Polar Bears: A Complete Guide to Their Biology and Behavior. Johns Hopkins University Press, Baltimore, Maryland.

Douglas, D.C, P.E. Reynolds, and E.B. Rhode. Editors. 2002. Arctic Refuge Coastal Plain Terrestrial Wildlife Research Summaries. USGS Biological Resources Division, Biological Science Report, USGS/BRD/BSR-2002-0001.

Doudna, J.A. and S.H. Sternberg. 2017. A Crack in Creation: Gene Editing and the Unthinkable Power to Control Evolution. Mariner Books, Houghton Mifflin Harcourt, Boston and New York.

Eggebo, S.L., K.F. Higgins, D.E. Naugle, and F.R. Quamen. 2003. Effects of CRP field age and cover type on ring-necked pheasants in eastern South Dakota. Wildlife Society Bulletin 31:779-785.

The Economist. 2007. Hail Linnaeus: Conservationists-and polar bears-should heed the lessons of economics. Economist.com. May 17, 2007.

Ehrlich, P.R. 2000. Human Natures. Island Press, Shearwater Books, Washington, D.C.

Federal Register. 1996. Policy Regarding the Recognition of Distinct Vertebrate Population Segments Under the Endangered Species Act. Vol. 61, Number 26, February 7, 1996, pages 4722-4725.

Federal Register. 2000. 12-Month Finding for a Petition to List the Black-Tailed Prairie Dog as Threatened. Vol. 65, No. 24, February 4, 2000, pages 5476-5488.

Federal Register. 2004. Finding for the Resubmitted Petition to List the Black-Tailed Prairie Dog as Threatened. Vol. 69, No. 159, August 18, 2004, pages 51217- 51226.

Federal Register. 2007. 12-Month petition finding and proposed rule to list the polar bear (*Ursus maritimus*) as threatened throughout its range. Vol. 72, No. 5, January 9, 2007, pages 1064-1099.

Federal Register. 2008. Endangered and Threatened Wildlife and Plants; Determination of Threatened Status for the Polar Bear (*Ursus maritimus*) Throughout Its Range; Final Rule. Vol. 73, No. 95, May 15, 2008, pages 28212-28303.

Federal Register. 2010. 12-Month Findings for Petitions to List the Greater Sage- Grouse (*Centrocercus urophasianus*) as Threatened or Endangered. Vol. 75, No. 55. March 23, 2010, pages 13910-14014.

Federal Register. 2014. Threatened Status for Gunnison Sage-Grouse. Vol. 79, No. 224, November 20, 2014, pages 69192-69310.

Federal Register. 2015. 12-Month Findings on a Petition to List Greater Sage-Grouse (*Centrocercus urophasianus*) as an Endangered or Threatened Species. Vol. 80, No. 191, October 2, 2015, pages 59858-59942.

Federal Register. 2016. 12 month finding on a petition to list the Alexander Archipelago wolf as an endangered or threatened species. Vol. 81, No. 3, January 6, 2016, pages 435-458.

Federal Register. 2017. Removing the Greater Yellowstone Ecosystem Population of Grizzly Bears from the Federal List of Endangered and Threatened Wildlife. Vol. 82, No. 125, June 30, 2017, pages 30502-30633.

Federal Register. 2019. Removing the Gray Wolf (*Canis lupus*) from the List of Endangered and Threatened Wildlife, Proposed Rule. Vol. 84, No. 51, March 15, 2019, pages 9648-9687.

Ferrara, P. 2013. The disgraceful episode of Lysenkoism brings us global warming theory. Forbes. http://www.forbes.com/sites/ peterferrara/2013/04/28/the-disgraceful-episode-of-lysenkoism-brings-us-global-warming-theory/2/.

Fitzsimmons, A.K. 1999. Defending Illusions: Federal Protection of Ecosystems. Rowman & Littlefield, Lanham, Maryland.

Francis, R.C. 2015. Domesticated: Evolution in a Man-Made World. W.W. Norton & Company, New York and London.

Fredrickson, R., P. Hedrick, R. Wayne, B. vonHoldt, and M. Phillips. 2015. Mexican wolves are a valid subspecies and an appropriate conservation target. Journal of Heredity 106:415-416.

Fukuyama, F. 1998. Women and the evolution of world politics. Foreign Affairs September/October 1998, pages 24-40.

Funk, W.C., E.D. Forsman, T.D. Mullins, and S.M. Haig. 2008. Introgression and dispersal among spotted owl subspecies. Evolutionary Applications pages 161-171. doi: 10.111/j.1752-4571.2007.00002.x

Futuyma, D.J. 1986. Evolutionary Biology. Sinauer Associates, Sunderland, Massachusetts.

Futuyma, D.J. 1995. Science on Trial: The Case for Evolution. Sinauer Associates, Sunderland Massachusetts.

Futuyma, D. J. and M. Kirkpatrick. 2017. Evolution 4th edition. Sinauer Associates, Sunderland, Massachusetts.

Gates, C.C., C.H. Freese, P.J.P. Gogan, and M. Kotzman. Editors. 2010. American bison, Status survey and conservation guidelines 2010. Gland, Switzerland. IUCN (International Union for Conservation of Nature and Natural Resources).

Geist, V. 1991. Phantom subspecies: the wood bison *Bison bison "athabascae"* Rhoads 1897, is not a valid taxon, but an ecotype. Arctic 44:283-300.

Geist, V. 1992. Endangered species and the law. Nature 357:274-276.

Gleiser, M. 2019. Atheism is inconsistent with the scientific method, prizewinning physicist says. Scientific American Interview with Lee Billings, March 20, 2019. https://www.scientificamerican. com/article/atheism-is-inconsistent-with-the-scientific-method-prizewinning-physicist-says/?utm_source=newsletter&utm_medium=email&utm_campaign=daily-digest&utm_content=link&utm_term=2019-03-21_featured-this-week&spMailingID=58798902&spUserID=Mzg0MjkwM-jQyODY3S0&spJobID=1602779071&spReportId=MTYwM-jc3OTA3MQS2.

Glynn, P. 1997. God: The Evidence. Forum, Imprint of Prima Publishing, Rocklin, California.

Good, T.P., R.S. Waples, and P. Adams. Editors. 2005. Updated status of federally listed ESUs of West Coast salmon and steelhead. U.S. Dept. Commerce, NOAA Technical Memo. NMFS-NWFSC-66. 598 pages.

Gould, S.J. 1977. Ontogeny and Phylogeny. Belknap Press, Cambridge, Massachusetts.

Grunwald, M. 2001. Departmental Differences Show Over ANWR Drilling. Washington Post, October 19, 2001.

Gutmann, S. 2000. The Kinder Gentler Military: Can America's Gender-Neutral Fighting Force Still Win Wars? Scribner, New York. The "great paradox" is in a quote is from former JAG lawyer Henry Hamilton on page 209.

Haig, S.M., R.S. Wagner, E.D. Forsman, and T.D. Mullins. 2001. Geographic variation and genetic structure in spotted owls. Conservation Genetics 2:25–40.

Haig, S.M., T.D. Mullins, and E.D. Forsman. 2004. Subspecific relationships and genetic structure in the spotted owl. Conservation Genetics 5:683-705.

Haig, S.M., E.A. Beever, S.M. Chambers, H.M. Draheim, B.D. Dugger, S. Dunham, and E. Elliott-Smith. 2006. Taxonomic considerations in listing subspecies under the US Endangered Species Act. Conservation Biology 20:1584–1594.

Halbert, N.D. and J.N. Derr. 2007. A comprehensive evaluation of cattle introgression into US federal bison herds. Journal of Heredity 98:1-12.

Haskell, S.P. and W.B. Ballard. 2004. Factors limiting productivity of the Central Arctic Caribou Herd of Alaska. Rangifer 24(2):71-78.

Haskell S.P. and W.B. Ballard. 2008. Annual re-habituation of calving caribou to oilfields in northern Alaska. Canadian Journal of Zoology 86:627-637.

Haskell, S.P., R.M. Nielson, W.B. Ballard, M.A. Cronin, and T.L. McDonald. 2006. Dynamic responses of calving caribou to oilfields in northern Alaska. Arctic 59:179-190.

Hedrick, P.W. 2009. Conservation genetics and North American bison (*Bison bison*). Journal of Heredity 100:411-420.

Horner, C.C. 2007. The Politically Incorrect Guide to Global Warming and Environmentalism. Regnery, Washington, D.C.

Hyland, W.G. 1990. The Cold War is Over. Random House, New York.

Itzin, C. Editor. 1992. Pornography: Women, Violence, and Civil Liberties A Radical New View. Oxford University Press, New York.

Johnson, P.E. 1993. Darwin on Trial. Intervarsity Press, Downers Grove, Illinois.

Joly, K., C. Nellemann, and I. Vistnes, 2006. A reevaluation of caribou distribution near an oilfield road on Alaska's North Slope: Wildlife Society Bulletin 34:866–869.

Joravsky, D. 1970. The Lysenko Affair. University of Chicago Press, Chicago.

Junger, S. 2016. Tribe: On Homecoming and Belonging. Twelve, Hachette Book Group, New York.

Kamarck, K.N. 2015. Women in Combat: Issues for Congress. December 3, 2015. Congressional Research Service 7-5700 www.crs.gov R42075.

Keegan, J. 1993. A History of Warfare. Vintage Books, Random House, New York.

Kelly, J.F. 2016. Department of Defense Press Briefing by General John Kelly January 8, 2016. https://dod.defense.gov/News/Transcripts/Transcript-View/Article/642104/

department-of-defense-press-briefing-by-general-kelly-in-the-pen-tagon-briefing/

Kesler, C.R. 2018. America's cold civil war. Imprimis 47(10):1-5, October 2018.

Kirchoff, M.D., T.M. Franklin, and J.W. Schoen. 1995. A model for science-based conservation advocacy. Wildlife Society Bulletin 23:358-364.

Lawrence, T.E. Undated. When asked, Why do men go to war? He answered Because the women are watching.

Levin, M.R. 2013. The Liberty Amendments, Restoring the American Republic. Threshold Editions, Simon & Schuster Inc. New York.

Lewontin, R. and R. Levins. 1976. The Problem of Lysenkoism. Pages 32-64 In: The Radicalisation of Science. H. Rose and S. Rose Editors. The MacMillan Press Ltd., London and Basingstoke.

Lomborg, B. 2001. The Skeptical Environmentalist: Measuring the Real State of the World. Cambridge University Press, Cambridge, United Kingdom.

Lorenz, K. 1971. Studies in Human and Animal Behavior Volume 2. Harvard University Press, Cambridge, Massachusetts.

Lueck, D. 2000. The Law and Politics of Federal Wildlife Preservation. Pages 61-119 In: Political Environmentalism: Going Behind the Green Curtain. T.L. Anderson Editor. Hoover Institution Press, Stanford University, Stanford, California.

Lyon, T.B. and W.N. Graves. 2014. The Real Wolf. L. Grosskopf and N. Morrison editors. Published by T.B. Lyon, and distributed by Farcountry Press, Helena, Montana.

MacDonald, S.O. and J.A. Cook. 2009. Recent Mammals of Alaska. University of Alaska Press, Fairbanks.

Maginnis, R.L. 2013. Deadly Consequences: How Cowards are Pushing Women into Combat. Regnery, Washington, D.C.

Maki, A.M. 1992. Of measured risks: the environmental impacts of the Prudhoe Bay, Alaska, oil field. Environmental Toxicology and Chemistry 12:1691-1707.

Malaney, J.L. and J.A. Cook. 2013. Using biogeographical history to inform conservation: the case of Preble's meadow jumping mouse Molecular Ecology 24(22):6000-6017.

Mann, C.C. and M.L. Plummer. 1995. Noah's Choice: The Future of Endangered Species. Alfred A. Knopf. New York.

Marris, E. 2007. The species and the specious. Nature 446:250-253.

Matthew 12:8. Other "Son of Man" verses include: Matthew 20:28; Matthew 16:13; Mark 2:10-11; Matthew 9:6; Luke 5:24; Mark 8:31; Luke 9:22; Acts 7:56; Mark 14:62; Matthew 26:64; Luke 22:69.

Mayr, E. 1954. Change of genetic environment and evolution. Pages 157-180 In: Evolution as a Process. J. Huxley, A. C. Hardy, and E. B. Ford Editors. Allen and Unwin, London.

Mayr, E. Editor. 1957. The Species Problem: A Symposium Presented at the Atlanta Meeting of the American Association for the Advancement of Science, December 28-29, 1955. Publication Number 50. The American Association for the Advancement of Science, Washington, D.C.

Mayr, E. 1970. Populations, species, and evolution. Belknap Press, Harvard University Press, Cambridge, Massachusetts.

Mayr, E. 1982. The Growth of Biological Thought. Belknap Press, Harvard University Press, Cambridge, Massachusetts.

Medvedev, Z. 1969. The Rise and Fall of T. D. Lysenko. Columbia University Press, New York.

Mitchell, B. 1989. Weak Link: the Feminization of the American Military. Regnery Gateway, Washington, D.C.

Mitchell, B. 1998. Women in the Military: Flirting with Disaster. Regnery, Washington, D.C.

Mojab, S. Editor. 2015. Marxism and Feminism. Zed Books, London.

Moritz, C. 1994. Applications of mitochondrial DNA analysis in conservation: a critical review. Molecular Ecology 3:401–411.

Morris, D. 1967. The Naked Ape: A Zoologist's Study of the Human Animal. Jonathan Cape Publishing, London.

Mukherjee, S. 2016. The Gene: An Intimate History. Scribner, New York.

Murray, W. 2017. America and the Future of War: The Past as Prologue. Hoover Institution Press, Stanford, California.

National Academies of Sciences, Engineering, and Medicine. 2019. Evaluating the taxonomic status of the Mexican gray wolf and the red wolf. The National Academies Press, Washington, D.C.

National Research Council. 1995. Science and the Endangered Species Act. The National Academies Press, Washington, D.C.

National Research Council. 2003. Cumulative environmental effects of oil and gas activities on Alaska's North Slope. The National Academies Press, Washington, D.C,

Noel, L.E., R.H. Pollard, W.B. Ballard, and M.A. Cronin. 1998. Activity and use of active gravel pads and tundra by caribou, *Rangifer tarandus granti*, within the Prudhoe Bay Oil Field, Alaska. Canadian Field Naturalist 112:400-409.

Noel, L.E., K.R. Parker, and M.A. Cronin. 2004. Caribou distribution near an oilfield road on Alaska's North Slope, 1978-2001. Wildlife Society Bulletin 32:757-771.

Noel, L.E., K.R. Parker, and M.A. Cronin. 2006. Response to Joly et al. (2006) A reevaluation of caribou distribution near an oilfield road on Alaska's North Slope. Wildlife Society Bulletin 34:870-873.

O'Connor, E. and V. Bergengruen. 2019. Military Doctors Told Them It Was Just "Female Problems." Weeks Later, They Were In The Hospital. Buzzfeed News. March 8, 2019, https://www.buzzfeednews.com/article/emaoconnor/woman-military-doctors-female-problems-health-care.

Oesterle, P., R. McLean, M. Dunbar, and L. Clark. 2005. Husbandry of wild caught greater sage-grouse. Wildlife Society Bulletin 33:1055-1061.

O'Gara, B.W. 2002. Taxonomy. Pages 3-65 In: North American Elk: Ecology and Management. D. E. Toweill and J. W. Thomas Editors. Smithsonian Institution Press, Washington D.C. and London.

Olsen, P.B. 2011a. The Nature of War Theory. U.S. Army War College Strategy Research Project Report for a Master of Strategic Studies Degree. (USAWC, 22March2}Ll).

Olsen, P.B. 2011b. Natural Selection and Nature of War. Small Wars Journal 14 November 2011. http://smallwarsjournal.com/blog/

natural-selection-and-nature-of-war; https://smallwarsjournal. com/blog/journal/docs-temp/757-olsen.pdf

Orning, E.K. 2014. Effect of predator removal on greater sage-grouse ecology in the Bighorn Basin conservation area of Wyoming. M.S. Thesis, Utah State University, Logan, Utah.

Patten, M.A. and K.F. Campbell. 2000. Typological thinking and the conservation of subspecies: the case of the San Clemente Island loggerhead shrike. Diversity and Distributions 6:177-188.

Patten, M.A. and J.V. Remsen Jr. 2017. Complementary roles of phenotype and genotype in subspecies delimitation. Journal of Heredity 108:462–464.

Pearce, J.M., P.L. Flint, T.C. Atwood, D.C. Douglas, L.G. Adams, H.E. Johnson, S.M. Arthur, and C.J. Latty. 2018. Summary of wildlife-related research on the coastal plain of the Arctic National Wildlife Refuge, Alaska, 2002–17. U.S. Geological Survey Open-File Report 2018–1003.

Polar Bear Specialist Group. 2019. The official website for the Polar Bear Specialist Group of the IUCN Species Survival Commission. http://pbsg.npolar.no/en/index.html.

Pollard, R.H., W.B. Ballard, L.E. Noel, and M.A. Cronin. 1996a. Summer distribution of caribou in the area of the Prudhoe Bay oil field, Alaska, 1990-1994. Canadian Field Naturalist 110:659-674.

Pollard, R.H., W.B. Ballard, L.E. Noel, and M.A. Cronin. 1996b. Parasitic insect abundance and microclimate of gravel pads and tundra within the Prudhoe Bay oil field, Alaska in relation to use by caribou. Canadian Field Naturalist 110:649-658.

Pombo, R. and J. Farah. 1996. This Land is our Land: How to End the War on Private Property. St. Martin's Press, New York.

Ramey, R.R., H-P. Liu, C.W. Epps, L.M. Carpenter, and J.D. Wehausen. 2005. Genetic relatedness of the Preble's meadow jumping mouse (*Zapus hudsonius preblei*) to nearby subspecies of *Z. hudsonius* as inferred from variation in cranial morphology, mitochondrial DNA, and microsatellite DNA: implications for taxonomy and conservation. Animal Conservation 8:329-346.

Ramey, R.R., H-P. Liu, C.W. Epps, L.M. Carpenter, and J.D. Wehausen. 2006. Response to Vignieri et al. (2006): should hypothesis testing or selective post hoc interpretation of results guide the allocation of conservation effort. Animal Conservation 9:244–247.

Ramey, R.R., J.D. Wehausen, H-P Liu, C.W. Epps, and L.M. Carpenter. 2007. How King et al. (2006) define an 'evolutionary distinct mouse subspecies': a response. Molecular Ecology 16:3518-3521.

Ranglack, D.H., L.K. Dobson, J.T. du Toit, and J. Derr. 2015. Genetic Analysis of the Henry Mountains Bison Herd. PLoS ONE 10(12): e0144239. doi:10.1371/journal.pone.0144239.

Rojas, M. 1992. The species problem and conservation: What are we protecting? Conservation Biology 6:170-178.

Roll-Hansen, N. 2005. The Lysenko Effect: The Politics of Science. Humanity Books, Imprint of Prometheus Books, Amherst, New York.

Rosenbaum, D.E. 2002. Senate Blocks Fuel Drilling in Alaska Wildlife Refuge. New York Times, April 19, 2002.

Roughgarden, J. 2006. Evolution and Christian Faith: Reflections of an Evolutionary Biologist. Island Press, Washington, D.C.

Sagan, C. 1977. The Dragons of Eden. Random House, New York.

Sanera, M. and J.S. Shaw. 1996. Facts not Fear: A Parent's Guide to Teaching Children about the Environment. Regnery, Washington, D.C.

Scalia, A. 2013. Opinion of Scalia, J. Supreme Court of the United States Nos. 11-338 and 11-347. Doug Decker, in his official capacity as Oregon State Forester et al. Petitioners 11-338 *v.* Northwest Environmental Defense Center Georgia Pacific West Inc., et al. Petitioners 11-347 *v.* Northwest Environmental Defense Center on writs of certiorari to the United States Court of Appeals for the Ninth Circuit [March20, 2013].

Schneider, L. 1986. Lysenkoism in China. Proceedings of the 1956 Qingdao Genetics Symposium. Tranlated by Q. Shizhen and L. Schneider. Chinese Law and Government, 1986 19(2). M. E. Sharpe, Armonk, New York.

Seck, H.H. 2018. Military changing body armor to accommodate women's hairstyles. Military.com 22 March 2018. https://www.military.com/daily-news/2018/03/22/military-changing-body-armor-accommodate-womens-hairstyles.html.

Serrano, L.F. 2014. Why Women Do Not Belong in the U.S. Infantry, Marine infantry isn't broken, it doesn't need to be "fixed". Marine Corps Gazette 98(9):36-40, September 2014.

Shaw, R.P. and Y. Wong. 1989. Genetic Seeds of Warfare. Unwin Hyman, Winchester, Massachusetts.

Simpson, G.G. 1961. Principles of Animal Taxonomy. Columbia University Press, New York.

Skalski, J.R., R.L. Townsend, L.L. McDonald, J.W. Kern, and J.J. Millspaugh. 2008. Type I errors linked to faulty statistical analyses of endangered subspecies classifications. Journal of Agricultural, Biological, and Environmental Statistics 13:199-220.

Smith, S. 2015. Women and Socialism: Class, Race and Capital. Haymarket Books, Chicago.

Solzhenitsyn, A.I. 1972. Nobel Prize Lecture. Solzhenitsyn received the Nobel Prize in Literature in 1970. The text of his Nobel speech appeared in 1972 and he received the Nobel insignia in person in Sweden in 1974.

Solzhenitsyn, A.I. 1974. Letter to the Soviet Leaders. Harper and Row, New York.

Soyfer, V.N. 1994. Lysenko and the Tragedy of Soviet Science. Rutgers University Press, New Brunswick, New Jersey.

Stirling, I. 2011. Polar Bears: The Natural History of a Threatened Species. Fitzhenry and Whiteside, Markham, Ontario, Canada.

Stirling, M.D. 2008. Green Gone Wild. Merril Press, Bellevue, Washington.

Strobel, L. 1998. The Case for Christ. Zondervan Publishing House, Grand Rapids, Michigan.

Studer, B. 2015. Communisme et féminisme (Communism and feminism). Clio, Women, Gender, History 41(1):139-152.

Sullivan, P.J., J.M. Acheson, P.L. Angermeier, T. Faast, J. Flemma, C.M. Jones, E.E. Knudsen, T.J. Minello, D.H. Secor, R. Wunderlich, and

B.A Zanetell. 2006. Defining and Implementing Best Available Science for Fisheries and Environmental Science, Policy, and Management. Fisheries 31(9):460-465.

Supreme Court of the United States 1984. Chevron U.S.A. v. Natural Resource Defense Council. June 25, 1984.

Supreme Court of the United States. 1997. Auer et al. v. Robbins et al. February 19, 1997.

Thomas, J.W. 2005. On Environmental Morality. Interview with Tim Findley, Range Magazine, Winter 2005.

Trans-Alaska Pipeline System (TAPS) Owners 2001. Environmental Report for the Trans-Alaska Pipeline System Right of Way Renewal. Trans-Alaska Pipeline System Owners, Anchorage, Alaska.

Truett, J.C. and S.R. Johnson. Editors. 2000. The Natural History of an Arctic Oil Field. Academic Press, San Diego, California.

U.S. Coast Guard. 2013. United States Coast Guard Arctic Strategy. United States Coast Guard Headquarters, CG-DCO-X, Washington, D.C. May, 2013.

U.S. Fish and Wildlife Service. 2008. Frequently asked questions about the Endangered Species Act listing of polar bears. http://www.fws. gov/home/feature/2008/polarbear012308/pdf/037257Polar-BearQAFINAL.pdf.

U.S. Fish and Wildlife Service. 2017. Polar bear Five Year Review: Summary and Evaluation. U.S. Fish and Wildlife Service, Marine Mammals Management, Anchorage, Alaska.

U.S. Geological Survey. 2007. USGS Science Strategy to Support U.S. Fish and Wildlife Service Polar Bear Listing Decision. Administrative Reports. U.S. Geological Survey. These reports are apparently not available on USGS websites. They are available at https://polarbearscience.com/2018/03/04/archive-of-2007-usgs-reports-supporting-2008-esa-listing-for-the-polar-bear/.

U.S. Geological Survey. 2019. Polar bear research. U.S. Geological Survey. https://www.usgs.gov/centers/asc/science/polar-bear-re-search?qt-science_center_objects=0#qt-science_center_objects.

U.S. Marine Corps. 2015. United States Marine Corps Ground Combat Element Integrated Task Force Research ONR Award

#N00014-14-1-0021 Final Report August 14, 2015. Neuromuscular Research Laboratory Department of Sports Medicine and Nutrition, University of Pittsburgh.

U.S. Navy. 2014. U.S. Navy Arctic Roadmap for 2014 to 2030. Department of the Navy, Chief of Naval Operations, Washington, D.C. February 20, 2014.

Van Creveld, M. 1991. The Transformation of War. The Free Press, Simon & Schuster, New York.

Van Creveld, M. 2001. Men, Women and War: Do Women Belong in the Front Line? Cassell & Co., London.

Van Loh, S.D. 2004. The Latest and Greatest Commerce Clause Challenges to the Endangered Species Act: Rancho Viejo and GDF Realty. Ecology Law Quarterly 31(3):459-485.

Vanzolini, P.E. 1992. Third world museums and biodiversity. Pages 185–198 In: Systematics, Ecology, and the Biodiversity Crisis. N. Eldredge Editor. Columbia University Press, New York.

Vayda, A.P. 1974. Warfare in ecological perspective. Annual Review of Ecology and Systematics 5:183-193.

Vermeire, L.T., R.K. Heitschmidt, P.S. Johnson, and B.F. Sowell. 2004. The Prairie Dog Story: Do We Have It Right? BioScience 54:689-695.

Webster. 1988. Webster's Ninth New Collegiate Dictionary. Merriam-Webster, Springfield, Massachusetts.

Weckworth, B.V., N.G. Dawson, S.L. Talbot, and J.A. Cook. 2015. Genetic distinctiveness of Alexander Archipelago wolves (*Canis lupus ligoni*): reply to Cronin et al. (2015). Journal of Heredity 106:412-414.

Wiley, E.O. 1978. The evolutionary species concept reconsidered. Systematic Zoology 27:17-26.

Wilson, E.O. 1975. Sociobiology. Harvard University Press, Cambridge, Massachusetts.

Wilson, E.O. 1978. On Human Nature. Harvard University Press, Cambridge, Massachusetts.

Wilson, E.O. 1994. Naturalist. Island Press, Shearwater Books, Washington, D.C.

Wilson, E.O. 2012. The Social Conquest of Earth. Liveright, New York, W.W. Norton & Company Ltd. London.

Wilson, E.O. and W.L. Brown. 1953. The subspecies concept and its taxonomic applications. Systematic Zoology 2:97–122.

Wrangham, R.W. and L. Glowacki. 2012. Intergroup aggression in chimpanzees and war in nomadic hunter-gatherers. Human Nature 23:5-29.

Young, N.M., T.D. Capellini, N.T. Roach, and Z. Alemseged. 2015. Fossil hominin shoulders support an African ape-like last common ancestor of humans and chimpanzees. Proceedings of the National Academy of Sciences 112(38):11829-11834.

Zink, R.M. 2004. The role of subspecies in obscuring avian biological diversity and misleading conservation policy. Proceedings of the Royal Society of London B 271:561-564.

Zink, R.M., G.F. Barrowclough, J.L. Atwood, and R.C. Blackwell. 2000. Genetics, taxonomy and conservation of the threatened California Gnatcatcher. Conservation Biology 14:1394-1405.

Zink, R.M., J.G. Groth, H. Vaquez-Miranda, and G.F. Barrowclough. 2014. Phylogeography of the California gnatcatcher (*Polioptila californica*) using multilocus DNA sequences and ecological niche modeling: implications for conservation. Auk 130:449-458.

INDEX

Wildlife, War, and God: Insights on Science and Government

Appendices

APPENDIX SPECIES AND SUBSPECIES SCIENTIFIC LATIN NAMES MENTIONED IN THE TEXT

Birds

Spotted owl	*Strix occidentalis*
Northern spotted owl	*Strix occidentalis caurina*
Mexican spotted owl	*Strix occidentalis lucida*
California spotted owl	*Strix occidentalis occidentalis*
Barred owl	*Strix varia*
Greater Sage-Grouse	*Centrocercus urophasianus*
Gunnison sage grouse	*Centrocercus minimus*
Coastal California gnatcatcher	*Polioptila californica californica*

Mammals

Black-tailed prairie dog	*Cynomys ludovicianus*
Preble's meadow jumping mouse	*Zapus hudsonius preblei*
Caribou	*Rangifer tarandus*
White-tailed deer	*Odocoileus virginianus*
Mule deer	*Odocoileus hemionus*
Grizzly or Brown Bear	*Ursus arctos*
Polar bear	*Ursus maritimus*
Black bear	*Ursus americanus*
Bison	*Bison bison*
Wood bison	*Bison bison athabascae*
Plains bison	*Bison bison bison*
Cattle	*Bos taurus*

Elk .... *Cervus elaphus*

Gray Wolf .... *Canis lupus*

Mexican wolf..*Canis lupus baileyi*

Red wolf .... *Canis rufus* **or** *Canis lupus rufus*

Eastern wolf .... *Canis lycaon* **or** *Canis lupus lycaon*

Alexander Archipelago wolf.............. *Canis lupus ligoni*

Plains wolf... *Canis lupus nubilus*

Coyote.. *Canis latrans*

Cheetah.. *Acinonyx jubatus*

Ringed seal ....*Pusa hispida*

Walrus.. *Odobenus rosmarus*

Ape... Family Hylobatidae, lesser apes; Family Hominidae, great apes and humans

Chimpanzee .... *Pan troglodytes*

Fish and amphibians

Salmon... *Oncorhynchus spp.*

Rainbow trout and steelhead.............. *Oncorhynchus mykiss*

Salamander .... **Order Urodela**

APPENDIX CHAPTER 4 POLAR BEARS, CLIMATE CHANGE, AND THE ESA

Included in this Appendix are:

1. Comments submitted by Matthew A. Cronin April 8, 2007 to the U.S. Fish and Wildlife Service on the proposed rule to list the polar bear as a threatened species under the ESA.
2. Comments submitted by Matthew A. Cronin October 21, 2007 to the U.S. Geological Survey on the USGS science strategy to support the proposed polar bear ESA listing.

1. Comments submitted by Matthew A. Cronin April 8, 2007 to the U.S. Fish and Wildlife Service on the proposed rule to list the polar bear as a threatened species under the ESA.

Matthew A. Cronin

8 April 2007.

Re: Federal Register 12-Month petition finding and proposed rule to list the polar bear (*Ursus maritimus*) as threatened throughout its range. Federal Register January 9, 2007, Vol. 72, No. 5, pages 1064-1099.

Please consider these comments on the Federal Register 12-Month petition finding and proposed rule to list the polar bear (*Ursus maritimus*) as threatened throughout its range (FWS 2007). The proposed rule was prepared by the U.S. Fish and Wildlife Service (FWS) and describes the analysis and decision that listing the polar bear as threatened under the Endangered Species Act (ESA) is warranted. Please note that I reviewed the version of the document in double-spaced manuscript format released prior to the Federal Register publication. My references to page numbers therefore will not reflect page numbers in the Federal Register version.

I oppose the listing of polar bears as a threatened species under the ESA. My comments below are offered as justification.

Non-scientific considerations

"I know no method to secure the repeal of bad or obnoxious laws so effective as their stringent execution". Ulysses S. Grant, Inaugural Address, 4 March 1869.

I offer the introductory quote by President Grant, and the comments below as constructive criticism. I know you must consider an overwhelming amount of scientific information, and thousands of comments from strident proponents and opponents of an ESA listing of polar bears. I hope my comments will contribute a scientific review of the Proposed Rule, and also stimulate thought and communication between you, the State of Alaska, and our fellow citizens on policy and management considerations. I wish you the best of luck in dealing with the issue before us.

You are certainly aware of the controversial nature of the ESA. My experience with working Americans in the natural resource industries and agriculture is that the ESA has become a bad and obnoxious law. It has been plagued by selective use of science to secure or maintain listings and has infringed on citizens' property rights (Pombo and Farah 1996). Consider that > 70% of the listed mammals in the U.S. are subspecies or Distinct Population Segments (DPS), not entire species (Cronin 2006). Also consider that the Proposed Rule under consideration is for a predicted future (not current) status of the polar bear. This species is not currently threatened and has healthy populations. The *Endangered Species* Act has been extensively applied to groups that are *not species*, and now to a group that is *not endangered*. I think you will agree that such extensive application, contrary to the very name of the act, is questionable at best. I urge you and your superiors in the Department of the Interior to reflect on these points. Also, recall your proper role in the federal government as defined in the U.S. Constitution which states:

"The powers not delegated to the United States by the Constitution, nor prohibited by it to the States, are reserved to the States respectively, or to the people." (10ᵗʰ Amendment to the U.S. Constitution)

The sweeping federal authority of the ESA is not defined in the Constitution, and fish and wildlife management is properly relegated to the states. I know that courts have allowed the federal government to use the ESA under the interstate commerce clause, which in my opinion is quite contrived. However, polar bears occur in only one state, Alaska, so even this justification is apparently invalid in the case at hand.

Scientific, policy, and editorial considerations

I am a scientist and have published several peer reviewed papers on polar bears (see References). Please consider my comments below as a peer review of the Proposed Rule and Status Review of the polar bears.

First, there is a question of scientific procedure. The entire system of review of science in the ESA process is closed. FWS (or National Marine Fisheries Service-NMFS for some species) either creates a petition or receives a petition from environmental groups to consider a species for ESA listing. FWS then reviews the petition and decides if it's warranted. FWS then prepares a Status Assessment/Review and a Proposed Rule and selects peer reviewers of them. FWS then reviews the peer reviews and produces a final Status Assessment and Proposed Rule. FWS then selects peer reviewers of Proposed Rule and solicits public comment. FWS then reviews the peer reviews and public comments and produces a final rule. FWS can dismiss or ignore public comments and peer review comments without accountability. The agency basically has the role of author, editor, and reviewer for their own documents. The potential for litigation challenges to FWS decisions in this process is slanted because courts give deference to agencies in assessing science and management information.

It is relevant that The Wildlife Society, whose majority membership is government agency biologists, has an official position statement that includes this statement:

"Oversight of scientific peer review should be vested in scientists and science managers within the agencies making decisions based on the science." (TWS 2006).

This policy would allow government agencies to control the science that is used and presented in policy and management documents and seems to mirror the closed peer review system for ESA that I described above.

Some of The Wildlife Society members also openly practice advocacy for wildlife management and conservation objectives over other legitimate resource objectives (TWS 1995). State and Federal wildlife agencies also form partnerships with environmental groups, who typically submit and support petitions for ESA listings. Finally, funding follows ESA listings, and agencies seek funding.

These points make clear the potential for bias and conflict of interest by agency biologists. This is an inherent problem with the ESA that needs attention from the Executive branch, Congress, and each of the 50 United States.

General comments

The proposed rule can be summarized as follows:

Because of climate change (i.e. global warming) arctic sea ice in summer is disappearing (by melting and removal by currents).

1. This will reduce the quality and quantity of polar bear habitat. Seals are the primary prey of polar bears, and the decline in sea ice will reduce seal habitat, and access to seals by polar bears.

2. Preliminary data/observations suggest that some polar bear populations may be experiencing negative impacts at present from decreased summer sea ice. This includes possible nutritional stress, lack of access to ice because of increased open water, intra-specific predation, mortality, and decline in some populations' numbers.

3. Considering 45 years as the "foreseeable future" it is concluded that polar bears will be threatened (i.e. they will be endangered with extinction) in that time frame.

4. A review of polar bear biology, and potential impacts from melting ice and other factors (e.g., hunting, disease, oil and gas development) is presented.

5. It's not clear how well the literature is covered on predicted climate change and sea ice conditions. This deserves intensive review and assessment by experts in these fields, as it is the basis of the entire issue.

6. In several places it is noted there is a lack of regulatory mechanisms to deal with the problem of warming and loss of sea ice (e.g., pages 90, 96, 116, 132, 133, and 135).

7. On page 149, Executive Order 13211 is addressed. This order of 18 May 2001 requires agencies to prepare "Statements of Energy Effects" if an agency action will affect energy supplies, distribution, or use. The FWS claims, (citing existing protections of the Marine Mammal Protection Act, MMPA), that this rule to list polar bears under the ESA is not a significant energy action and

no Statement of Energy Effects is required. This decision seems based on the premise that the listing will not affect oil and gas operations in the Alaskan arctic. However, it is relevant that the FWS says on page 141 that the Denali Commission may be involved in this listing under Section 7 because of potential funding of fuel and power generation projects, some of which presumably will not be within the range of polar bears.

8. On page 150, the FWS declared that no Environmental Assessment (EA) or Environmental Impact Statement (EIS) under the National Environmental Policy Act (NEPA) is needed, citing a previous notice (Federal Register 25, 1983, 48 FR 49244). This implies that the rule to list polar bears under the ESA is not a significant federal action.

9. On page 146, the FWS states they believe that "unauthorized destruction or alteration of the denning, feeding, resting, or habitats used for travel that actually kills or injures individuals (*sic*) polar bears by significantly impairing their essential behavioral patterns including breeding, feeding, or sheltering..." could potentially result in a violation of ESA section 9 and associated regulations (50 CFR17.31) with regard to polar bears.

A broad assessment of the Proposed Rule shows that there is no presentation or testing of hypotheses, despite the fact that the relevant assessment of polar bears and their habitat is predictive in nature. Simple assumptions that warming and ice disappearance will continue and that polar bears will decline across their entire range to the point of near extinction are made. It is reasonable to conclude that if sea ice declines significantly or entirely, polar bear populations will decline. However no quantitative analysis or models of population numbers (or prey abundance) are made. The analysis is speculative.

Points 7, 8, 9, and 10 above seem very important considering the summary statement (page 135): "...we find that the polar bear is likely within the foreseeable future ... to become an endangered species throughout all or a significant portion of its range based on threats to the species, including loss of habitat caused by sea ice recession and lack of effective regulatory mechanisms to address the recession of sea ice. Therefore we propose to list the polar bear as threatened."

Regarding point 7, if the proposed rule is needed because no regulatory mechanism currently exists, then the polar bear ESA listing must be such a regulatory mechanism to protect the species. Because the species is threatened by arctic sea ice loss, which is attributed to climate change, which is attributed to human greenhouse gas emissions, then it must be greenhouse gas emissions that are to be regulated. If greenhouse gas emissions are causing the melting ice, and melting ice is altering or destroying polar bear habitat and impairing their feeding and sheltering behaviors, then greenhouse gas emissions become a violation of section 9 and associated regulations. Because government authority is restricted to the U.S., regulation of American greenhouse gas emissions will be the target of a regulatory mechanism (i.e., the polar bear ESA listing). This will affect many parts of the American economy and is clearly a significant federal action. Note that an EIS under NEPA may require an analysis of the economic impacts of the action, in addition to an analysis of the environmental impacts. This is critical to determine the cost/benefit of regulation of greenhouse gases under the ESA.

However, the recent Supreme Court decision that Environmental Protection Agency (EPA) can regulate greenhouse gases thought to be responsible for global warming means that there is now a regulatory mechanism to address the threat to polar bear habitat. An ESA listing would be redundant and divert attention from the broad problem to focus on one species. I suggest the Proposed Rule be withdrawn because of this.

Regarding Executive Order 13211 and NEPA requirements, it is clear that the proposed rule will affect much more than arctic Alaskan oil and gas activity, including energy issues nationwide. For example, independent of the polar bear issue, the Attorney General of the State of California has sued automobile manufacturers over global warming. Major energy projects (e.g. the proposed Alaska natural gas pipeline) could be challenged on the grounds that they will contribute greenhouse gases to the atmosphere and harm an endangered species' habitat. The polar bear ESA ruling will give additional grounds for such lawsuits and regulatory actions, and clearly will be a significant energy action and a significant federal action. Ignoring Executive Order 13211 and NEPA requirements is not appropriate in my opinion.

Distinct Population Segments (DPS)

On page 30 of the pre-federal register version of the proposed rule, FWS notes they found the entire species meets the definition of a threatened species under the Act, and did not consider the alternative of assessing different DPS. However, on page 5, comments are solicited on whether any of the populations may qualify as DPS.

There is also a section on DPS in the Status Review (page 34) that is seriously deficient. There is a short discussion of bear territories followed by one sentence noting that telemetry data indicate spatial segregation of stocks in different regions. The Status Review doesn't even contain the term DPS in the text, nor does it present other relevant cases for comparison (e.g. grizzly bear DPS in the lower 48 states). Notably, the Status Review doesn't even mention the 19 putative subpopulations (mentioned in the Proposed Rule) as potential DPS.

Because polar bears are highly mobile, with relatively low level of genetic differentiation among subpopulations (Paetkau et al. 1999, Cronin et al. 2006), genetic differentiation is not likely to be useful or meaningful in designation of DPS or other management units (Cronin 1993, 2007a). This is particularly likely because of changes in movements and distribution that will result if the predicted changes in sea ice habitat occur.

Note the testimony of Richard Glenn given in Barrow in March 2007 (RDC 2007) in which he describes movements of bears on sea ice and under their own power. This results of interchange between stocks (i.e., subpopulations) that must be taken into account when assessing changes to a population. As a local resident and hunter, Glenn has intimate knowledge of this for the subpopulations in the Beaufort Sea, Chukchi Sea, and arctic Canada. His local knowledge is supported by modern science (Cronin et al. 2006). The point is that it is important to consider immigration and emigration when assessing causes of change in populations' numbers.

The Proposed Rule (pre-federal register version, pages 22-23) discusses the 19 different polar bear populations, and that they are discrete enough to manage each independently. However, the claim in the proposed rule that correspondence between genetic and movement data reinforces this is not correct. See Cronin et al. (2006), in which genetic and movement data do not clearly correspond. The Proposed Rule also says the populations are "relatively discrete populations". Such an uncertain level of discreteness is common for contiguous con-specific populations. It is important to note the lack of complete discreteness of these populations because of the potential for DPS to be identified in the future.

The proposed rule also says the population boundaries were developed from decades of intensive scientific studies. However, many of the populations' status are unknown so the intensity of these studies is questionable. This should be duly acknowledged and corrected.

In addition, it should be stated, if not emphasized, that subspecies and DPS categories used under the ESA are subjective and arbitrary (Cronin 2006) and the criteria to designate them are so general that populations in most geographic areas would qualify. All that is required is "marked" separation, such as physical (geographic) separation. It is up to FWS to decide what constitutes DPS and they have essentially full discretion to do so. It is relevant that FWS and NMFS developed the criteria for DPS, have authority to designate DPS, and respond to or dismiss peer review and public comment on DPS. This is a serious scientific deficiency in previous ESA assessments in which subjectively defined subspecies and DPS are presented as definitive groups (Cronin 2006, 2007a, 2007b). This was the case for the western Alaska stock of sea otters, which were declared a DPS with subjective criteria (Cronin 2004). I attached the comments submitted on the sea otter case for your information. Please consider these general points about DPS when considering polar bears, and my previous comments on the closed system of review of science for the ESA.

POLAR BEAR POPULATION NUMBERS

Pages 22-27 of the Proposed Rule indicate that polar bears are not currently threatened with being endangered with extinction. It is only speculation considering the prediction of arctic sea ice melting. The following are the numbers of animals in the 19 polar bear populations in the world identified in the Proposed Rule (from an IUCN report):

East Greenland, number unknown, trend unknown
Barents Sea, number = 3000, trend unknown
Kara Sea, number unknown, trend unknown
Laptev Sea, number = 800-1200, trend unknown
Chukchi Sea, number = 2000, trend unknown
Southern Beaufort Sea (primary Alaska population), number = 1500, trend declining (but not statistically significant).
Northern Beaufort Sea, number = 1200, trend stable
Viscount-Melville, number = 215, trend increasing (following overharvest)
Norwegian Bay, number = 190, trend declining
Lancaster Sound, number = 2541, trend stable
M'Clintock Channel, number = 284, trend increasing (following overharvest)

Gulf of Boothia, number = 1523, trend stable

Foxe Basin, number = 2197, trend stable

West Hudson Bay, number = 935, trend declining

South Hudson Bay, number = 1000, trend stable

Kane Basin, number =164, trend declining

Baffin Bay, number = 2074, trend declining

Davis Strait, number = 1650, trend unknown

Arctic Basin, number unknown, trend unknown.

Number populations declining = 5

Number populations stable = 5

Number populations increasing = 2

Number population unknown = 7

Total number of populations =19.

As I understand it, these population estimates are fairly rough because of the difficult nature of gathering information in the Arctic. In any event, there are 5 populations that are declining in number. Note that the West Hudson Bay population declined, the south Hudson Bay population is stable. Also, note that the decline in the Southern Beaufort Sea population is not statistically significant (Regehr et al. 2006).). Also, note the testimony of Patterk Netser, Minister of Environment, Government of Nunavut given in Washington, D.C. March 2007 in which he reveals that the surveys of the Western Hudson Bay subpopulation did not include the entire range, so the estimate of numbers of animals (and hence a subpopulation decline) is not scientifically definitive (RDC 2007). Please do not ignore such information from those who live in, and have intimate knowledge of, the polar bears' habitat.

It is important to note that 37%, 7 of 19 populations' status is unknown. It is premature to declare the species threatened with extinction with so little information on the species' status. There are other cases where FWS found a species to be threatened or endangered, only to change the decision when additional data was gathered (e.g., sage grouse, black-tailed prairie dog, California tidewater goby).

Consider the following estimates of past polar bear numbers worldwide:

> The proposed rule (FWS 2006): 20,000-25,000 in 2006,
> Amstrup (2003, citing Lunn et al. 2002) 21,500-25,000 in 2002,
> Scribner et al. (1997, citing Wiig et al. 1995) 21,000-28,000 in 1995,
> Servheen (1990, citing Larsen 1984) 25,000 in 1984-1989,
> Servheen (1990, citing Maksimov and Sokolov 1965, Cowan 1972)
> 8,000-10,000 in 1965-1970.

It appears that polar bears have increased two-three times in number worldwide since 1965-1970, and many populations are stable, increasing, or of unknown status. Polar bears are not currently threatened with extinction. The potential threat is dependent on the future sea ice conditions.

The Proposed Rule did not have any numerical estimates of the extent of the predicted decline of polar bear numbers. This is a serious deficiency because the proposed rule found that polar bears will be threatened with extinction in 45 years. However, the same group that provided the population estimates used in the Proposed Rule (the IUCN) did provide a numerical estimate. Regarding the prediction of future declines of polar bears because of loss of sea ice habitat, the IUCN Report (Aars et al. 2006, page 61) notes that the upgrading polar bears to "Vulnerable" status is "based on the likelihood of an overall decline in the size of the total population of more than 30% within the next 35 to 50 years. The principal cause of this decline is climatic warming and its consequent negative affects (*sic*) on the sea ice habitat of polar bears."

The IUCN prediction of a decline in the size of the total population of more than 30% within the next 35 to 50 years does not result in a population that is threatened with extinction (e.g., 30% of 20,000-25,000 is 6,000-7,500, leaving minimally 14,000-17,500 surviving bears). The Congressional Research Service notes that an estimate of the minimum viable total population estimate for polar bears is 4,961 individuals (Buck 2007, citing Reed et al. 2003).

These projected changes to polar bear populations do not constitute a threat of extinction.

Survival during previous warming periods
Predictions of future polar bear population status should also consider that polar bears survived previous warming periods, including periods 8,000-11,000 years ago and 1000 years ago (Buck 2007). This is noted in the Proposed Rule (page 72) but no insights are offered regarding the predicted warming of concern to extant populations. It is well known that many species went extinct about 12,000 years ago (termed the megafaunal extinction) but this didn't include polar bears. Interestingly, cheetahs also escaped the megafaunal extinctions of 12,000 years ago but suffered a severe loss of genetic variation, presumably from a population bottleneck at that time (O'Brien and Johnson 2005). Genetic variation appears not to be reduced in polar bears compared to their ancestral species (brown bears, Paetkau et al. 1997, 1998, 1999, Cronin et al. 2005, 2006). This is not hard evidence there was not a population bottleneck of polar bears during previous warming periods, but it is suggestive that there were not population reductions severe enough to reduce genetic variation as in cheetahs. FWS willingness to predict future polar bear population status should be complemented with a backward look at past populations and habitat conditions.

Future Summer Arctic Sea Ice Conditions
On pages 31- 36 of the Proposed Rule, an overview of arctic sea ice change is presented, referring to climate models and data. Little or no

reference is made to any study not predicting rapid warming and arctic sea ice loss.

The Proposed Rule and Status Assessment apparently uses selected science that presents a maximum loss in the shortest time of Arctic summer sea ice. For example, Zhang and Walsh (2006) show model output with a wide range of the extent and timing of Arctic summer sea ice. Consideration of alternative hypotheses is needed to employ the best available science in this enterprise. Also, Zhang and Walsh note the considerable uncertainty in the models. This should be openly acknowledged and emphasized in the Proposed Rule.

In addition, it needs to be clearly and repeatedly acknowledged that the sea ice predictions are based on models with various assumptions and parameters that may or may not be realistic. Model results should be treated as hypotheses, testable only with data collected in the future. Therefore consideration of all legitimate hypotheses regarding the timing and extent of sea ice change, and the factors causing the change is the proper way to do a scientific assessment of the issue. For example, a proper way to assess the issue is to first establish null hypothesis:

Ho: there will be no difference in Arctic summer sea ice conditions between 1979 (or some average considered appropriate) and 2052 (45 years from now).

Alternative hypotheses can be entertained, for example:

H1: there will be a 90% reduction of Arctic summer sea ice between 2007 and 2052
H2: there will be a 10% reduction of Arctic summer sea ice between 2007 and 2052

These and other hypotheses can be tested over time, but for now models predicting the ice conditions can be employed. The Service made a feeble attempt at this by referencing a few models and accepting the extreme case of ice loss from those models. Appropriate science would be to consider all models and different outcomes from each. A range of predicted ice conditions depending on the model, and their assumptions and data put into them, would then be generated.

An important peer reviewed paper is In Press (Dyck et al. 2007), and should be consulted for a good scientific appraisal of the issue of climate change and polar bears.

Additional comments

On page 11, it is stated that genetic research had **confirmed** that polar bears evolved from grizzly (brown) bears 250,000-300,000 years ago, citing Cronin et al. 1991a:2290. This is incorrect. First, Cronin et al. noted that the **fossil record** (not genetic data) **suggests** (not confirms) that polar and brown bears diverged less than 300,000 years ago, and that this is a rough estimate because the fossil record of bears is not complete. In addition, Cronin et al. explicitly stated that **use of the genetic data** (mitochondrial DNA data in this case) for dating divergences of taxa (e.g. species) **was unwarranted** in the case of the bears. The proposed rule also cites Talbot and Shields 1996:574) on this issue. Talbot and Shields did use genetic data to estimate divergence time of the brown and polar bear lineages. However, they placed the divergence of brown and polar bears at 300,000-400,000 years ago, not 250,000-300,000 years ago. (My emphasis in **bold**.)

Also, on page 8 of the Status Review Yu et al. (2004) is cited for mtDNA data supporting recent evolution of polar bears from grizzly bears. Yu et al. do not present mtDNA data (they analyzed other genes) and do not mention polar bears (they assessed brown bears and Asiatic black bears).

This misreporting of the literature is problematic, particularly the belief that the divergence estimates were confirmed. It shows a willingness to accept estimates as hard fact by imprecise reading of the literature and limited understanding of science outside the area of wildlife biology. If this type of deficient review of science was also done for the climate and sea ice issues (which is likely outside the expertise of the authors of the proposed rule), the entire predicted response of polar bears to climate change is questionable and should be thoroughly reviewed by climate and oceanography experts and all reasonable models considered.

There are also statements inappropriate for a modern scientific work. On page 12, it is stated that polar bear adaptations include "small ears **for** reduced surface area; ...feet with ..."suction cups" on the underside **for** increased traction on ice..." On page 13, it is stated that "Polar bears evolved **to utilize** the Arctic sea ice niche..." On page 129 it is stated that "...polar bears have **evolved to occur** throughout the ice-covered waters..."

The words in bold print imply that traits evolved for a purpose. Gould and Lewontin (1979) and Gould (2002) described the problems with creating "just so" stories to explain adaptations. Our understanding of evolution is that random mutations (without purpose) occur, and natural selection will favor those that enhance reproduction and survival. First, mutations conferring smaller ears or suction cups on feet would have occurred without regard for environmental conditions. Second, we do not **know** that these traits are due to additive genetic variance that was selected. It is a reasonable inference that variance in ear and foot morphology may be due to genetic variance, but many morphological traits co-vary according to general growth patterns. That is, small ears may have evolved as a by-product of a general change in growth pattern of polar bears as they lived on sea ice. The small ears did not necessarily evolve for reduced surface area. In addition, there may be environmental effects on growth patterns, and traits that vary between species and subspecies may be due to environment or heritability. We may observe that the ears of polar bears have less surface area and the feet have more traction than those of relatives (say grizzly bear ears and feet). It is more appropriate to state adaptive explanations simply as the function they serve, without inference about directionality. One should say, "small ears **with** reduced surface area compared to related species", "feet **with** increased traction on ice compared to related species", and "evolved **in** the sea ice niche". This may seem like trivial semantics, and for understanding the issue at hand it is semantic. However, precise language is a hallmark of science. The important point is that there is a systemic problem with the proposed rule and many other ESA assessments. Wildlife biologists have assumed the role of assessing science outside their expertise. In the case of the proposed rule, this may apply to climatology and oceanography as well as evolutionary biology. (My emphasis in **bold.**)

Page 12, the statement about cross-breeding of polar bears and grizzly bears in the wild needs a citation.

Pages 17-18, reproduction is discussed, including the 3-year breeding interval of females. Cub mortality is discussed elsewhere. It should be discussed that if cub mortality results in subsequent estrus in the mother and reproduction resumes then the reproductive interval will be shortened. This could be an important parameter in population modeling.

Page 20, the Proposed Rule states that there were anomalous heavy ice conditions in the southern Beaufort Sea in the mid 1970's and mid 1980's, each lasting 3 years. The heavy ice caused decline in productivity of seals and natality and survival of polar bears. This seems inconsistent with the general points that sea ice is decreasing and sea ice is good habitat for seals and polar bears. There was also apparently extensive winter ice in 2005-2006 in the Bering Sea extending farther south than normal. This warrants consideration while the sea ice projections are being assessed.

Page 21, It seems that polar bears in western and southern Hudson Bay have adapted (behaviorally) to lower ice levels than elsewhere. Also denning occurs in different habitats in different areas. This could be important and used in modeling polar bears worldwide response to reduced ice.

Pages 28-29, the time frame for considering "foreseeable future" was calculated. FWS used the IUCN SSC Red List criteria of 10 years or three generations. A generation defined by IUCN is age of sexual maturity (5 years) plus 50% of the length or the lifetime reproductive period (20 years). One generation for polar bears is then calculated as 15 years, and 3 generations as 45 years.

These numbers appear to be for females because male polar bears may not begin breeding until 7 years of age. This point should be clarified. Also, some females may breed at 4 years of age (S. Amstrup, M. Cronin, unpublished genetic data) so the age of sexual maturity may need revision. No modeling of the population dynamics, including assessment of changes in carrying capacity, for bears over this time frame is provided. Such modeling, under different realistic future ice conditions should be done for an adequate assessment. I believe population modeling was done for the Steller sea lion after its designation as endangered. As noted above, predictions of future polar bear populations were made by the IUCN.

In many places the future status of polar bears is stated definitively when it should be stated with the qualification that it is a prediction. These are basically speculations:

> page 36 "Observed and predicted changes in sea ice cover, characteristics and   timing **have** profound effects on polar bears."

Page 37 "This increased rate and extent of sea ice movements **requires** additional efforts and energy expenditure for bears to maintain their position...

Page 44 "...adaptive behaviors of using terrestrial habitat instead of sea ice **will not** offset energy losses from decreased seal consumption, and nutritional stress **will** result."

Page 48 "While it is possible that reduced ice cover along with increased open and warmer water will enhance primary productivity of sea prey items, and thus seal productivity, ultimately such a regime **will** negatively impact polar bears."

Page 51 "Thus, a decrease in ringed seal abundance and availability **would** result in a concomitant decline in polar bear populations."

Page 56 "In the fall of 2004 four polar bears were observed **to have drowned** while attempting to swim between shore and distant pack ice in the Beaufort Sea." And page 59 "...four polar bears drowned in open water..." I believe the drowning was a hypothesized cause of death, and the animals were only observed from the air so the cause of death was not determined.

Page 68-69 "This **will** result in fragmentation of habitat...these factors **will** negatively impact polar bears by increasing the energetic demands...redistributing ...populations ... and increasing levels of negative bear-human interactions." "These factors **will,** in turn, **result** in the reduced ...condition of polar bears which **leads** to population-level demographic declines through reduction of survival and recruitment rates." "The ultimate effect of these ...factors...will be that polar bear populations **will decline or continue to decline.**"

Note Table 1 on pages 69-70 identifies "predicted   changes" while the text states things definitively as noted above.

Page 70 "The southerly populations of ...**will** be affected earliest." Earlier melt periods...**will** result in lengthened...use of land and increased...fasting, **resulting**   in decreased ... condition for bears in these populations." "Other populations ...**will**, or **are** currently, experiencing initial effects of changes in sea ice".

Page 73 "Changes in the timing of sea ice formation ... **will** pose increasing risk to polar bears...and ultimately **affect** all polar bear populations and **threaten** the species throughout all or a significant portion of its range in the foreseeable future." We find that polar bear populations...**are threatened** by ...changes in ...sea ice..."

Page 130 "The eventual effect **would be** that polar bear populations **will** continue   to decline."

Page 134 "...these factors...**will** result in declines or continued declines for all   populations."...within the foreseeable future... all populations **will** be either   directly or indirectly impacted."

Page 37, it is stated that "Polar bears are inefficient moving on land... when walking." Presumably they are inefficient walking on ice also. This points warrants discussion because of polar bears extensive use of sea ice, and because the polar bear is the most highly mobile non-aquatic mammal. It seems inconsistent that the bears walk extensively on ice but are inefficient when walking.

Page 40, the reference to Schliebe (unpublished data) regarding a trend of increasing use of coastal areas by polar bears in the fall open water period in the southern Beaufort Sea, should be made available on a website or other access.

Page 43, it is observed that a majority of native elders and senior hunters in Nunavut, Canada (83%) believes the population of polar bears has increased. The FWS says the increase was attributed to more bears seen near communities, cabins, and camps, and hunters encountering bear sign in areas not previously used by bears, without identifying who attributed this. That is, the claim that the native observations were due to more bears near human activity needs documentation. It is then noted that "Some people interviewed" said this could reflect a change in behavior rather than an increase in the population. The proportion comprising the "some people" should be given. The entire results of the native interviews should be cited and made available as a source of traditional ecological knowledge (TEK). This issue is also noted on pages 81-84.

There is an important related point on page 110. The FWS says that the Territory of Nunavut may place greater significance on indigenous knowledge than on scientific data and analysis. The FWS position on indigenous knowledge needs clear and consistent definition. Alaska natives will provide input on issues regarding polar bears, and how FWS will treat it should be known beforehand, not after the fact.

Page 44, it is stated that polar bears have been observed using terrestrial food items such as snow geese, but it is stated below in the same paragraph that polar bears are not known to hunt snow geese.

Page 47, it is noted that declines in seal numbers have resulted in "marked declines" in polar bear populations. The marked declines should be defined in terms of time and numbers so future populations can be quantitatively modeled.

Page 48, has a summary of the issue: "Thus, major declines in sea ice habitats as projected will likely result in a decline in polar bear abundance over time due to reduced prey availability (Derocher et al. 2004:167)."

This appears to be the overall appraisal of FWS also. However, declines in abundance are common in wildlife populations and does not necessarily constitute a threat of extinction. It may be a reasonable hypothesis that polar bears may decline so dramatically as to be threatened with extinction, but it is simply a hypothesis, and should not be presented as a conclusion.

The uncertainty of the magnitude and extent of impact of melting ice on polar bears is supported on page 51 where it is stated "The potential effects of sea ice change on population size are difficult to quantify." This is followed by a discussion of physical condition, reproduction, and survival, characteristics associated with elevated risks of extinction for other species. This statement accentuates the hypothetical nature of the impacts on polar bears, especially an impact severe and extensive enough to cause extinction.

Page 55, it is stated that a significant decline may occur in the southern Beaufort Sea population if the trends in sea ice continue as predicted and this trend will occur in the 45 year "foreseeable future". Again, the future condition of sea ice and polar bear populations are hypotheses, and should be presented as such.

Page 63, reference is made to Fishbach et al. (2005) findings of increased numbers of dens on land and fewer on pack ice. It would be worth discussing other variables that would affect proportions of dens on land and sea ice (e.g., den detection methods, access to human-killed whale carcasses).

Page 64, the issue of thermal properties of snow changing and dens is discussed. An example of captive cubs held without a den at -45F and dying within two days is given. This should be put in perspective of the likelihood of cubs not having a den in the wild. It seems obvious that the young of any warm blooded species would not survive such extremes and this point is of questionable relevance to the discussion.

Page 65, the statement about levels of oil and gas activity in different countries ne eds citations.

Page 66, an NRC (2003) report is summarized, and increased displacement from oil and gas facilities is assumed. This may not be supported by data. I believe that the experience at BP's Northstar offshore drilling island in the Beaufort Sea has been that polar bears visit the island frequently. Also, recall that increased numbers of bears around human activity are noted on page 43 of the Proposed Rule. The evidence for displacement needs critical review.

Page 67, indicates that current regulations such as Letters of Agreement (LOA) have minimized impacts of oil and gas on polar bears. Only two bear mortalities have occurred in Alaska, compared to 33 in Canada, as a result of human conflict.

Pages 49 and 58, extends concerns over sea ice conditions to walrus. "These observations indicate that the Pacific walrus population may be ill-adapted to rapid seasonal sea ice retreat off Arctic continental shelves". For several years the wildlife agencies have advocated the need and superiority of "ecosystem management" over "single species management". The concerns over walrus, is an indication that changes in the sea ice environment will affect the entire ecosystem, not only polar bears. It could be argued that the Proposed Rule (and the entire ESA for that matter) is for an issue that should be dealt with at the ecosystem level (changing of the entire Arctic Ocean ice conditions) not a single species.

Page 69, it is stated: "Not all populations will be affected evenly in the level, rate, and timing of impact, but within the foreseeable future, it is predicted that all populations will be either directly or indirectly impacted." The basis of this prediction and the extent of the impact really need to be quantified. To just say the ice will melt and the bears will decline to near extinction is not acceptable for such an important issue.

Page 71, it is discussed that some feel that "if sea ice disappeared altogether, polar bears would become extinct". This, of course, is the important unknown, how much and how fast will sea ice decline. It also indicates the tenuous nature of the proposed rule finding that the species is actually threatened with endangerment of extinction. The finding is based to a large extent on feelings, not quantitative science.

Pages 71-72, it is noted that the opinion that polar bears are likely to go extinct is not shared by all polar bear biologists (without citations). This documentation is needed. It is also noted that polar bears have survived previous warming periods. It is stated that "The precise impacts of these warming periods on polar bears and the Arctic sea ice habitat are unknown." This sentence should also apply to FWS' analysis in the Proposed Rule of future ice conditions, of which the impacts on the bears' populations are hypothetical and without quantification. It is not consistent to (appropriately) acknowledge the lack of knowledge the past, while stating with certainty that future warming will threaten the species with extinction.

On page 73 (and pages 86, 130, 133), the impacts from commercial, scientific, and other purposes is discussed. No mention of research and management impacts, including mortality, is given. This needs to be assessed. For example captive cubs being held without a den at -45F with resulting death needs explanation. Mortalities from research and management-related capture with drugs, traps, etc. should also be quantified. Dyck et al. (2007) discuss the potential for research and management activities to negatively affect polar bears. The importance of this issue is apparent as an Environmental Impact Statement (EIS) is being prepared assessing the impacts of research on Steller sea lions.

On pages 79-80, the MMPA is discussed. Polar bears are not listed as depleted under the MMPA, so it seems inappropriate for them to be considered threatened under the ESA. It is also noted that Canada has managed polar bears so as to be maintained at a sustainable level.

On page 81, harvest of polar bears is discussed: "Five populations have no estimate of potential risk from over harvest, since adequate demographic information necessary to conduct a population viability analysis and risk assessment are not available". Information to conduct risk assessment is not available for harvest analysis, but the prediction of ice melting is enough to predict extinction? The speculative nature of future sea ice conditions and polar bears' response to them is apparent in this comparative case.

On page 128, the finding is stated while claiming that all information available was considered. This is a dubious claim. For example, climate models that do not predict a severe a loss of sea ice were not considered in the Proposed Rule. In addition, the IUCN prediction of a 30% population reduction was not used. Also, 12 peer reviewers were noted, who were experts in polar bear ecology and other relevant fields. The selection, qualifications, and potential conflicts of interest of the peer reviewers should be made available.

On page 135, the summary statement is made: "...we find that the polar bear is likely within the foreseeable future ... to become an endangered species throughout all or a significant portion of its range based on threats to the species, including loss of habitat caused by sea ice recession and lack of effective regulatory mechanisms to address the recession of sea ice. Therefore we propose to list the polar bear as threatened." As noted above, the recent Supreme Court decision that the Environmental Protection Agency (EPA) should regulate greenhouse gases provides a regulatory mechanism and the proposed rule is unwarranted.

On page 139, the term "essential habitat" is used when referring to polar sea ice. Because this term has a regulatory meaning for certain fish, an alternative term should be used (e.g., important habitat).

On page 141, the Denali Commission is noted as a group that may get involved with section 7 of the ESA because it funds fuel and power projects. This implies that projects dealing with energy in general (i.e., not in polar bear habitat) will be under ESA regulatory scrutiny. The thought that the FWS and its partner environmental groups will be regulating the energy industry is profound in its potential effects on the economy, private property rights, and other issues (e.g. Pombo and Farah 1996).

Pages 145-146, the issue of incidental take is discussed. FWS needs to identify and define all direct and incidental take before the proposed rule is allowed to proceed. FWS states that unauthorized destruction or alteration of habitats are possible violations of section 9. Conceivably, any greenhouse gas emission could be considered as contributing to habitat alteration and therefore constitute a take.

On page 148, comments are solicited on how to make the proposed rule easier to understand. I suggest that the proposed rule should clearly separate fact from speculation, especially the repeated statements that Arctic sea ice will melt and polar bears may go extinct without any biological modeling or formulation of hypotheses. The proposed rule could also include a section on potential bias and advocacy on the part of FWS, with acknowledgement that wildlife agencies practice wildlife conservation advocacy (see Cronin 2007, TWS 1995).

Thank you for your consideration.

REFERENCES

Aars, J., N. J. Lunn, and A. E. Derocher (Compilers and editors). 2006. Polar bears. Proceedings of the 14th working meeting of the IUCN/SSC polar bear specialist group, 10-24 June 2005, Seattle Washington, USA. Occasional Paper of the IUCN species Survival commission No. 32.

Amstrup, S.C., G.W. Garner, M.A. Cronin, and J.C. Patton. 1993. Sex identification of polar bears from blood and tissue samples. Canadian Journal of Zoology, 71(11):2174-2177.

Amstrup, S.C. 2003. Polar bear. Chapter 27 *in* Wild Mammals of North America: biology, management and conservation. Edited by G.A. Feldhamer, B.C. Thompson, and J.A. Chapman. John Hopkins University Press. Baltimore. Pages 587-610.

Buck, E. H. 2007. Polar bears: Proposed listing under the Endangered Species Act. Congressional Research Service (CRS) Report for Congress. 30 March 2007

Cowan, I.M. 1972. The Status and Conservation of Bears of the World-1970. International Conference on Bear Research and Management 2:343-367.

Cronin, M.A., S.C. Amstrup, G.W. Garner, and E.R. Vyse. 1991a. Interspecific and intraspecific mitochondrial DNA variation in North American bears (*Ursus*). Canadian Journal of Zoology 69:2985-2992.

Cronin, M.A., D.A. Palmisciano, E.R. Vyse, and D.G. Cameron. 1991b. Mitochondrial DNA in wildlife forensic science: species identification of tissues. Wildlife Society Bulletin 19:94-105.

Cronin, M. A. 1993. Mitochondrial DNA in wildlife taxonomy and conservation biology: cautionary notes. Wildlife Society Bulletin 21:339-348.

Cronin, M.A. 2004 Review of science related to the Proposed ESA listing of the Southwest Alaska Stock of Sea Otters (Federal Register/ Vol. 69, No. 28 February 11, 2004/ Proposed Rule. Department of the Interior, Fish and Wildlife Service, 50 CFR Part 17 RIN 1018-AI44). Unpublished Report, Matthew A. Cronin, University of Alaska, 14 December 2004.

Cronin, M.A., R. Shideler, L. Waits, and R.J. Nelson. 2005. Genetic variation and relatedness in grizzly bears (*Ursus arctos*) in the Prudhoe Bay region and adjacent areas in northern Alaska. Ursus 16:70-84.

Cronin, M. A., S. C. Amstrup, and K. T. Scribner. 2006. Microsatellite DNA and mitochondrial DNA variation in polar bears in the Beaufort and Chukchi seas, Alaska. Canadian Journal of Zoology 84:655-660.

Cronin, M.A. (2006). A Proposal to eliminate redundant terminology for intra-species groups. Wildlife Society Bulletin 34: 237-241.

Cronin, M.A. 2007a. The Preble's meadow jumping mouse (*Zapus hudsonius preblei*): subjective subspecies, advocacy, and management. Animal Conservation. In Press.

Cronin, M. A. 2007b. EIEIO (Evolutionarily interesting and ecologically important organisms). Range Magazine Spring 2007: 24-25.

Derocher, A.E., N.J. Lunn, and I. Stirling. 2004. Polar bears in a warming climate. Integrative and Comparative Biology 44:163-176.

Dyck, M. G., W. Soon, R. K. Baydack, D. R. Legates, S. Baliunas, T. F. Ball, and L. O. Hancock. 2007. Polar bears of western Hudson Bay and climate change: Are warming spring arir temperatures the "ultimate" survival control factor? Ecological Complexity In Press.

Fishbach, A.S., S.C. Amstrup, and D. Douglas. 2005. Identifying polar bear denning behavior by satellite radio telemetry reveals changes in northern Alaska denning distribution. Poster presented at 17th Biennial Marine Mammal Conference, San Diego, California, USA. December 2005.

FWS (U.S. Fish and Wildlife Service) 2007. 12-Month petition finding and proposed rule to list the polar bear (*Ursus maritimus*) as threatened throughout its range. Federal Register January 9, 2007, Vol. 72, No. 5, pages 1064-1099.

Gould, S.J. and R.C. Lewontin. 1979. The spandrels of San Marco and the Panglossian paradigm: a critique of the adaptationist programme. Proceedings of the Royal Society of London B. 205:581-595.

Gould, S. J. 2002. The Structure of Evolutionary Theory. The Belknap Press of Harvard University Press. Cambridge, Massachusetts. 1433 pages.

Larson, T. 1984. We've saved the ice bear. Int. Wildl. 14:4-11.

Lomborg B. 2001. The Skeptical Environmentalist. Cambridge University Press, Cambridge, United Kingdom. 515 pages.

Lunn, N.J., S. Schiebe, and E.W. Born, eds. 2002. Polar bears. Proceedings of the 13th Working Meeting of the IUCN/SSC polar bear specialist group. Occasional Paper of the International Union for Conservation of Nature Species Survival Commission No. 26, Gland, Switzerland.

Maksimov, L.A. and V.K. Sololov. 1965. Polar bear: distribution and status of stocks; problems of conservation and research. Pages 39-43 *in* Proceedings First International Meeting on Polar Bear. University of Alaska.

National Research Council (NRC). 2003. Cumulative Environmental Effects of Oil and Gas Activities on Alaska's North Slope. National Academies Press, Washington, D.C. 288 pages.

O'Brien, S. J. and W. E. Johnson 2005. Big cat genomics. Annual Review of Genomics and Human Genetics. 6:407-429.

Pombo, R. and J. Farah. 1996. This Land is Our Land. St. Martin's Press. New York. 225 pages.

Reed, D. H. et al. 2003. Estimates of minimum viable population sizes for vertebrates and factors influencing these estimates. Biological Conservation 113:23-34.

Regehr, R.V., S.C. Amstrup, and I. Stirling. 2006. Polar bear population status in the southern Beaufort Sea. U.S. Geological Survey Open-File Report 2006-1337, 20p.

Schliebe, S.L., T.J. Evans, and M.A. Cronin. 1999. Use of genetics to verify sex of harvested polar bears: Management implications. Wildlife Society Bulletin 27:592-597.

Scribner, K.T., G.W. Garner, S.C. Amstrup, and M.A. Cronin. 1997. Population genetic studies of the polar bear: A summary of available data and interpretation of results. Pages 185-196 in A.E. Dizon, S.J. Chivers, and W.F. Perrin eds. Molecular Genetics of Marine Mammals. Society for Marine Mammalogy Special Publication 3.

Servheen, C. 1990. The Status and Conservation of the Bears of the World. International Conference on Bear Research and Management Monograph Series No. 2. 32 pages.

Talbot, S. L. and G. F. Shields 1996. A Phylogeny of the bears (Ursidae) from complete sequences of three mitochondrial genes. Molecular Phylogenetics and Evolution 5:567-575.

TWS (The Wildlife Society) 1995. Wildlife Society Bulletin. Vol. 23 number 3.

TWS (The Wildlife Society). 2006. TWS position statement: Scientific peer review of agency decision processes. The Wildlifer Issue number 336, May-June 2006: 6, 16.

Wiig A., E.W. born, and G.W. Garner, eds. 1995. Proceedings of the 11th Working Meeting of the IUCN/SSC Polar Bear Specialist Group, 25-27 January 1993, Copenhagen Denmark, IUCN, Gland, Switzerland. 192 pages.

Wiig, O., A. E. Derocher, M. A. Cronin, and J. U. Skaare. 1998. Female pseudohermaphrodite polar bears at Svalbard. The Journal of Wildlife Diseases. 34:792-796.

Zhang, X., and J. E. Walsh, 2006: Toward a seasonally ice-covered Arctic Ocean: Scenarios from the IPCC AR4 model simulations. *J. Climate, 19*, 1730-1747.

2. Comments submitted by Matthew A. Cronin October 21, 2007 to the U.S. Geological Survey on the USGS science strategy to support the proposed polar bear ESA listing.
Comments on U.S. Geological Survey reports (USGS science strategy to support U.S. Fish and Wildlife Service polar bear listing decision) on polar bears in support of the ESA listing issue.

Matthew A. Cronin, Ph.D.
21 October 2007
The comments in this review pertain to the U.S. Geological Survey (USGS) reports on polar bears related to the Endangered Species Act (http://www.usgs.gov/newsroom/special/polar_bears/, Federal Register 20 September 2007 (72 FR 53749):USGS science strategy to support U.S. Fish and Wildlife Service polar bear listing decision, and

Federal Register 5 October 2007, Volume 72, Number 193, Extension of comment period).

Please note that I previously provided comments to the U.S. Fish and Wildlife Service (FWS) on the proposed rule that are relevant to the reports reviewed here (letter from M. A. Cronin to FWS, 8 April 2007, regarding Federal Register 12-Month petition finding and proposed rule to list the polar bear (*Ursus maritimus*) as threatened throughout its range. Federal Register January 9, 2007, Vol. 72, No. 5, pages 1064-1099).

Relation of the USGS reports to the proposed ESA rule

It is apparent that the proposed rule and status review were premature and inadequate if additional studies were needed to assess the basic premises of current and future status of polar bear populations. I think there should be accountability for this, as many in state government, industry, and citizens have had to spend large amounts of time down loading, printing, reviewing, and commenting twice on one issue because of the inadequacy of the proposed rule. From a scientific perspective, if the basic premises of a work are not well supported it doesn't pass peer-review. This is the case for the proposed ESA rule. In the case of management, the proposed rule is also inadequate. Consider a private-sector comparison. Imagine if a timber company submitted a management plan to harvest federal timber, and upon submission announced that they had serious information deficiencies and were going to rush some new studies, but expected to maintain the original schedule. They would be told to submit a complete plan when it is ready, not in pieces with admitted deficiencies.

I reviewed the nine USGS reports regarding polar bears as if I was peer-reviewing scientific papers. As such, I am brief, and perhaps blunt, in the interest of conciseness. I do not intend personal criticism and I do not profess expertise in all of the topics reported. Consider that there are 17 different individuals who are co-authors of the nine reports, with a wide range of expertise and experience. My review reflects that of an informed biologist, not an expert in climate, sea ice, or polar bears. As such, I expect some of my questions and insights will be shared by other biologists and policy makers.

I hope my comments and those of others are seriously considered as constructive criticism. My experience with the FWS and National Marine Fisheries Service (NMFS) is that comments on Endangered Species Act (ESA) listings are ignored or dismissed without serious consideration or potential for appeal. I hope USGS involvement can prevent this in the case of the polar bears.

Also, I ask questions that may be answered in the reports, but were not clear to me in reading the large amount of material in a short amount of time. For example, the estimates and measurements of polar bear survival are critical to many of the analyses predicting future population numbers, but it was not clear to me how survival was estimated.

I commend the USGS scientists who had to conduct this complex work in a small amount of time. Because of the great importance of the polar bear ESA issue I hope the reports can be improved with review and synthesis.

Consequences of basing ESA decisions on predictive models

The status of polar bears worldwide is currently good, including a wide distribution and healthy population sizes. FWS' proposed threat of extinction depends entirely on the predictions of future habitat loss from changes in climate and sea ice conditions. If polar bears are found to be threatened with extinction for ESA purposes, I offer the following declaration for use by FWS and the Secretary of the Interior, that includes an appropriate qualifier:

> *Polar bears are threatened with extinction if the climate, ice, and biological models upon which we relied are accurate in their predictions. Polar bears are not threatened with extinction at this time, and the threatened finding depends entirely on if the climate and ice habitat changes occur as modeled.*

I am not contesting the legitimacy of the climate, ice, or ensuing biological modeling (but see my review of the reports below). I am simply pointing out my understanding of the reality of the situation and the consequences of basing important decisions on predictive demographic and climate models.

Consider the precedent established if the USGS reports are used to support the declaration of a species as threatened with extinction based on predictive modeling. There is no limit to the number of species, subspecies, and distinct population segments (DPS) that can be subjected to climate or other models and predicted to be threatened in the future. There will no longer be a scientific standard for empirical data and information, only predictions derived from models. The potential for new ESA listing petitions to expand greatly is apparent. Previously, the use of subjectively-defined subspecies and DPS made the number of potential ESA candidates essentially limitless. If predictive scenarios of threats of extinction are allowed, as with the polar bears, findings of threatened or endangered status will also be limitless. This warrants reflection on the proper role of the federal government in what is traditionally state jurisdiction of resource management. It seems more appropriate for the federal government to deal with climate change as an issue by itself (basically an air pollution issue), and not allow proliferation of ESA cases to astronomical levels.

General comments on the USGS reports
I offer an observation regarding the title of the project "USGS science strategy *to support* U.S. Fish and Wildlife Service polar bear listing

decision" (my italics). This implies the studies were done to support the listing decision (i.e., FWS has already determined a listing was warranted). I think the reports should be described as research on polar bears relevant to the proposed ESA listing, for use by the FWS and American citizens.

It is telling that the USGS reports predict significant declines in polar bear populations. This is not surprising because they rely on the same climate/sea ice models that predict warming and loss of sea ice. However, the reports openly acknowledge considerable uncertainty in the climate/sea ice and polar bear population models. I therefore feel that the discussion and conclusion sections are speculative about a negative future status of polar bears. I think it would be more appropriate for the USGS reports to state results, and prominently note that the likelihood of their occurrence depends entirely on the independent sea ice and climate models (which after all are not the results of the USGS reports). I offer the scientific insight that because of uncertainty in the climate and sea ice models, many of the polar bear results are hypotheses, not conclusions.

I suggest that an approach more explicitly employing hypotheses should be employed by USGS and acknowledged by the FWS in their use of this information. The reports could be revised with the following format:

1. Present explicit hypotheses and how they will be tested.
2. Describe models, assumptions, and analyses so they can be replicated by others.
3. Make available the data used in the analyses.
4. Present the modeling results.
5. In the case of predictive results, present them as hypotheses about polar bear populations' future status, to be tested with data in the future. Identify what data will be needed to test them and over what time scales.

6. Note that if the sea ice model predictions are realized, these modeling results for polar bears might occur. In my opinion the uncertainty of the climate/sea ice models makes it inappropriate for biologists to express certainty about the polar bear models.

Numbers 2 and 4 have been done in the current reports. Numbers 1, 3, and 5 can make the reports more scientifically rigorous.

Consider that the pending ESA listing decision for polar bears will potentially be based on predictions, when the world population of polar bears has not demonstrably declined. It seems inappropriate and unprecedented to list a species as threatened with extinction when it currently is not. Overall, I think presenting the results as hypotheses rather than conclusions is the most scientifically valid way to make this situation clear to policy makers and the public.

My reviews of the individual reports follow.

Review of Amstrup et al. Forecasting the range-wide status of polar bears at selected times in the 21st century.

The Bayesian network (BN) model is described in the abstract as a prototype and throughout the report as in need of review, refinement, and addition of additional expert judgments. It does not seem to meet the standard of being peer reviewed and may not be appropriate for use on the ESA issue at this time. The model based on carrying capacity is simpler, easier to understand, and may provide useful information at present.

Regardless, the BN and carrying capacity modeling efforts rely largely on the sea ice models, which in turn rely on climate models. Therefore, the complex biological modeling can be boiled down to understanding that if the ice declines, polar bears will decline. The key is whether the ice models will be accurate in their predictions.

The assumption that there is a linear relationship between sea ice extent and polar bear numbers (page 19) seems a reasonable estimation of carrying capacity, but (as noted on page 11) needs quantification regarding prey, density effects, and other factors. Also, 50% and 15% ice cover, and 127 ice-free days, are used as thresholds in the other studies, and how such quantities relate to a linear relationship seems relevant. It wasn't clear to me if carrying capacity is being defined primarily by summer sea ice conditions, and not ice during other seasons (e.g., page 26: "Our carrying capacity model therefore does not account of these seasonal aspects of sea ice change"). If bears can withstand fasting during the summer open water season (as in Hudson Bay) or if winter conditions affect carrying capacity, then year round conditions seem important and there may be too much emphasis on summer ice conditions.

I question the exclusive use of satellite data to assess sea ice. Do these data accurately depict ice as it is used by polar bears? It seems that simple aircraft monitoring of ice conditions a few times during the summer and fall could augment the satellite data with more biologically-relevant data. Also, on page 9 "ice cover" is defined as areas with ≥50% ice concentration. This means some areas classified as not being ice covered have up to 49% ice in them. Durner et al. apparently use ice cover as low as 15% in some analyses.

There are assumptions about bears' avoidance of human developments and impacts from oil and gas development (page 13) that need to be documented with citations. Limited impact of existing oil and gas development is noted with citations on page 31.

The conclusion (page 36) that "Polar bear populations in the polar basin divergent and seasonal ice ecoregions will most likely be extirpated by mid century" seems inappropriate as a conclusion. It is a prediction, based on a polar bear model that was based on an ice model that was based on climate models. I think it should be presented as a testable hypothesis, not a conclusion.

Review of Bergen et al.: Predicting movements of female polar bears between summer sea ice foraging habitats and terrestrial denning habitats of Alaska in the 21ˢᵗ century: proposed methodology and pilot assessment.

This study is a pilot assessment and makes the point that increased ice-free water may require bears to travel longer distances from ice habitats to terrestrial den habitats. Like the other assessments, it is dependent on the sea ice models that are dependent on the climate models.

Again, I question the exclusive use of satellite data to assess sea ice. Do these data accurately depict ice as it is used by polar bears? It seems simple aircraft monitoring of ice conditions a few times during the summer and fall could augment the satellite data with more biologically relevant data. Also, on page 9 "good sea ice habitat" is defined as areas with ≥50% ice concentration. This means some areas classified as not being good sea ice habitat have up to 49% ice in them. The habitat values of these areas are therefore not clear. Durner et al. apparently use ice cover as low as 15% in some analyses.

The use of 1979 as the first year of sea ice data appears to be because that is the first year satellite data were available. This should be augmented with earlier records to assess the longer term patterns of sea ice extent and duration.

It is not clear why if a greater proportion of dens have occurred on land, it is also a future necessity. It is also possible that if there are longer periods of open water, bears will den more frequently on sea ice, and not travel to terrestrial areas. I would think there is literature on other species changing their migratory patterns if habitats change.

Review of Obbard et al.: Polar bear population status in southern Hudson Bay, Canada.

This analysis found no significant change (but a slight decline) in survival and a non-significant increase in population size between 1984-86 and 2003-05. I think the authors speculated too much toward an assumed negative impact. The authors state (page 14):

"changes in ...sea ice... have yet not resulted in ...changes in survival or to a... reduction in population size in the SH population...".

The use of the word "yet" means they are anticipating the population will have these changes in the future and is therefore speculation.

In the abstract it is stated that they estimated a 5% decline in survival of sub-adult and adult females and a 7% decline in survival for sub-adult and adult males between 1984-86 and 2003-05, and noted the overlap of confidence intervals prevented them from unequivocally concluding that this was not a chance occurrence (i.e., there was not a statistically defensible decline in survival). They used the word "decline", with the qualification that it was not significant. This is accurate.

In the case of population numbers it is stated in the abstract that the population size appears to be unchanged and the population estimates were 641 in 1984-1986 and 681 in 2003-2005 with the confidence intervals indicating this is a non-statistically significant difference. Note there was an increase of 40 in the population size (= 6% of the 1984-86 estimate).

To be consistent they could have said there was a 6% increase in population size but it was not statistically significant. This subtle wording (decline in survival and unchanged population size) stresses a negative impact on the bears, when both parameters are statistically non-significant. Statistics should be applied consistently.

The authors continue with comparison of the status of the western Hudson Bay population to speculate (page 14): "If the body condition of polar bears in SH continues to decline, effects on reproduction will become evident. In addition, the declines in survival shown in the present study will become more pronounced. The net result will be a subsequent decline in the size of this (SH) population." These predictions expressing certainty (the words "will become or "will be"), should be framed as hypotheses to be tested with data in the future. Also, this statement claims that a decline in survival is shown in this study. The statistical non-significance of the decline makes this statement inappropriate.

Review of Stirling et al.: Polar bear population status in the northern Beaufort Sea.

In the abstract and page 9 the results of this study are summarized:

"Models that allowed associations between annual variation in (polar bear) survival, habitat, or relative seal abundance variables were not, in general, supported by the data."

"Currently the NB polar bear population appears to be stable, probably because ice conditions remain suitable..."

Results include population estimates:
1972-75: 745
1985-87: 867
2004-06: 980

These estimates show an increasing trend, but were not statistically significant and there were large confidence intervals around these point estimates. On page 10 it is noted: "estimates of abundance were remarkably similar through the 1970s, 1980s, and 2000s."

However, the population apparently increased 31% (235 bears/745 bears) between 1972-75 and 2004-06 if the point estimates are used. On page 13 it is noted that the estimate for 2004-06 may be conservative because of a strong possibility of an underestimate in 2006, and the estimates of 1100-1200 in 2004 and 2005 may more accurately reflect the current number of bears in the NB. I think they should present the 2004-06 estimates as 980-1200 to be more accurate.

Regarding survival (page 9): "Survival in the 1970s was comparable to that in the 2000s, ...lower in the 1980s than in the 2000s, and ...higher in the 1990s than in the 2000s."

And on page 11: "...estimated survival rates of bears of each age and sex class between sampling periods...were remarkably consistent."

The NB population has not shown statistically significant decreases in numbers or survival from the 1970s to 2006. Numbers actually appear to have increased. It seems that if the FWS proposed ESA rule were correct, there should be declines in survival and numbers in this population, as FWS claims the entire species is threatened with extinction. The analysis of the NB population does not support the (unstated) hypothesis of FWS that the entire species will decline to near extinction. Again, this is dependent on climate and ice models independent of this report.

I suggest that hypotheses should be formulated regarding the NB population. As of 2006, there seems to be no support for the hypothesis that ice loss is occurring or impacting the NB population. The fact that the current models don't show a decline of ice or numbers of bears points to the FWS' reliance on predictions in finding that polar bears are threatened with extinction.

The authors noted significant mixing of northern and southern segments of the population. They also assessed emigration, referring to (temporary or permanent) movement out of the study area. These observations will be useful in addressing management units and distinct population segments (DPS) in the future.

The authors noted that no bears died during capture in this study of 1031 marked bears. The number of research-related mortalities should be reported for the other studies done by USGS and the impact of disturbance and mortality assessed. Note that NMFS has recently been ordered to do an Environmental Impact Statement (EIS) on impacts on Steller's sea lions from their research and management activities.

On page 6, the decadal oscillation was included as a covariate in the analyses. This seems to greatly complicate the assumptions of the ice and climate models. Perhaps the decadal oscillation is included in the ice and climate models already.

On page 11 it is noted that in 1985-87 and the mid-1970s ringed seal productivity declined (with potential impact on polar bear survival) because of very heavy ice. The inconsistency between the primary problem of less ice, and this observation of a problem with heavy ice, is an apparent contradiction that should be clarified. It appears that there is an optimal amount of ice (summer and winter) and there can be too much or too little.

For the modelers' consideration I note that in the abstract and on page 9 it is stated:

"Models that allowed associations between annual variation in (polar bear) survival, habitat, or relative seal abundance variables were not, in general, supported by the data.", and on page 12: "The ice covariate in our analysis was not significant".

Does this mean there is no relationship between ice extent and polar bear survival? This seems to be in conflict with the results of the other USGS reports.

Review of DeWeaver: Uncertainty in climate model projections of Arctic sea ice decline: an evaluation relevant to polar bears.
The author states in the abstract: "A key point in the discussion is that the inherent climate sensitivity of sea ice leads inevitably to uncertainty in simulations of sea ice decline."

Developing and presenting models with such uncertainty is a legitimate scientific endeavor. However, I submit it is not appropriate as the primary information in a significant management action such as an ESA listing. Inherent uncertainty in sea ice prediction, leads to inherent uncertainty in polar bears' response. The entire issue has considerable uncertainty and use in an ESA listing decision is questionable.

As I suggested in comments on the proposed rule, model results should be treated as hypotheses, testable with data collected in the future. Therefore consideration of all legitimate hypotheses regarding the timing and extent of sea ice change, and the factors causing the change, is a proper way to do a scientific assessment of the issue. For example, a null hypothesis could be developed:

Ho: there will be no difference in Arctic summer sea ice conditions between 1979 (or some average considered appropriate) and 2052 (45 years from now).

Alternative hypotheses can be entertained, for example:
H1: there will be a 90% reduction of Arctic summer sea ice between 2007 and 2052
H2: there will be a 10% reduction of Arctic summer sea ice between 2007 and 2052

These and other hypotheses can be tested over time with observational data.

In this report sea ice extent is defined as the area with $\geq 50\%$ ice cover. It is important to know if maps depicting the ice margin excludes areas with $< 50\%$ ice cover. If so, they are not really accurate. Bergen et al. appropriately note in their map (Fig. 1d) the ice shown is $\geq 50\%$ concentration. If there is habitat value (e.g., resting areas for long distance swimming to avoid drowning) for polar bears when ice is $< 50\%$ concentration, it should be accounted for in some manner.

Review of Regehr et al. Polar bears in the southern Beaufort Sea I: Survival and breeding in relation to sea ice conditions, 2001-2006. It is not clear why the study was limited to 2001-2006. There are apparently satellite ice data from 1979 onward, and the polar bears in the SB have been studied since 1967 (page 18). It seems that a simple plotting of population numbers, survival, and cub recruitment, vs. ice extent and duration from 1979-2006 would be informative.

It wasn't clear to me in this study or that of Hunter et al. how survival was estimated and included in the models. Also, on page 4, the symbol for survival is apparently missing in the text.

On page 4 it is stated that female polar bears generally first breed at 5 years old. However, some 4-year old females probably breed in the SBS. Also, sub-adult males are identified as 2, 3, and 4 years old, but some males as young as 3 years old may breed in the SBS. These observations, and inference of reproduction by individuals, and females breeding <2.5 years apart, have been made with field and genetic data (Cronin, Amstrup, Talbot, Sage, unpublished manuscript), and may be useful in the analyses done in this report. Also, genetic analysis can provide estimates of emigration and immigration between populations (e.g., Cronin et al. 2006, Can. J. Zool. 84:655) that could be used in demographic analyses and the assessment of temporary emigration (page 11).

As in the other USGS reports, sea ice extent is defined as the area with $\geq$50% ice cover. Assessment of habitat value for conditions with 0%-49% ice cover is needed. Also, Durner et al. apparently use a threshold of 15% ice cover for some analyses.

627 bears were caught in this study (page 12). The potential impact of capture and mortalities should be reported, as did Stirling et al.

On page 12 it is stated: "The mean duration of the ice-free period in 2006 was 16.7 days longer than in 1979, although the trend was not statistically significant". This is Regehr et al.'s result, but then they reference Hunter et al. (2007) who used different statistics to "suggest" a tendency for a higher frequency of years with a longer ice-free season. This reference to another study with indefinite results should be in the discussion, not the results section. Also, Hunter et al. only found 21% of the years between 1979 and 2005 were "bad" ice years. Note also that Rode et al. found no significant trend in the sea ice cover between 1982 and 2006 in the SBS. Trends in the ice-free period in the SBS appear equivocal considering all of the USGS reports.

On page 13 it is noted that survival varies little with the covariate *ice* until there are 127 days of ice-free conditions. 127 days may be a 'threshold" at which survival is negatively impacted. This may allow formation of informed hypotheses regarding polar bears' response to ice extent and duration.

On page 14 the temporary emigration rate is given as 0.2. This apparently means that bears leave the study area about 20% of the time. This is consistent with an apparent high rate of gene flow between the SBS and Chukchi Sea populations (Cronin et al. 2006, Can. J. Zool. 84:655). Temporary emigration is also discussed on page 17. This reflects the non-independence of study areas or population (or subpopulation depending on one's terminology preference) ranges in a species with high mobility and gene flow. This may be relevant to the issue of distinct population segments (DPS) noted in the proposed rule.

In several places a relationship between sea ice and polar bear survival is noted as a result of the study:

On page 16 in the discussion it is stated that anecdotal reports of emaciated, drowned, and cannibalized polar bears in the last 3 years, compared with few or none previously, is "consistent with the statistical relationships between sea ice and vital rates evidenced in our analysis".

On page 17 it is stated that "survival varied as a function of the covariate *ice...*",

On page 18 (Conclusions) it is stated that "...our analyses show evidence for an association between declining sea ice and reduced survival.", and "...the ....study in the SB region from 2001-2006 established a relationship between declining sea ice and decreased survival, and

On page 18 (Summary) it is stated "The declines in survival and breeding were associated with increases in the duration of the ice-free period..."

The model results are quite complex and I may not have understood them all, but these statements appear to overstate the results. It seems that they refer primarily to the relationship of ice-free periods > 127 days and survival (page 13) and the observed ice-free periods >127 days in 2004 and 2005 (page 12). It is stated that "Within the range of observed values, survival varied little with the covariate *ice* up to a threshold value of about 127 ice-free days."

If this is the case and there is not a general relationship of ice and survival, vital rates, and breeding, then the relationship could be restated explicitly (i.e., ice free periods > 127 days are statistically associated with survival). I think it is important if there is not a general relationship between ice and polar bear survival, vital rates, and breeding, and if the relationship is primarily with survival and >127 day ice free periods.

The authors also note (page 18, Conclusions) "It is very difficult to quantify demographic trends on the basis of 5 yearly intervals from 2001-2006..." This casts reasonable caution on their results and makes uncertain any general relationships between ice and survival.

This limitation and the apparent restriction of a significant relationship to survival and ice-free periods > 127 days seem like important qualifications.

It is questionable whether the discussion on sea ice forecasts and the preliminary 2007 ice extent should be in the conclusions (page 18). The conclusions should be restricted to those from the study itself. Putting the results in the context of the other literature is appropriate for the discussion section.

Review of Hunter et al. Polar bears in the southern Beaufort Sea II: Demography and population growth in relation to sea ice conditions.
This study used much of the information and assumptions in Regehr et al.

I think there are problems with the use of "good" years of 2001-2003 and "bad" years of 2004-2005 (with regard to sea ice conditions). It seems critical that between 1979 and 2005, there were 6 of 28 (21%) years with "bad" ice conditions and 22 of 28 years, 79% with good ice conditions. On page 9 it is shown that during the good years (2001-2003) the population does well and during the bad years (2004-2005) it does not. Given these opposite trends in sequential years, is there really justification to predict the future 45 years with any certainty?

As I note for the Rode et al. report, the data appear to show an ice-bear relationship on an annual basis, not a cumulative effect. I recognize the climate and ice models incorporate other considerations (such as positive feedback from loss of ice albedo etc.), but I think the analyses may be missing the "forest for the trees" by reliance on models, instead of data that is in hand. Indeed, on page 12 the authors note that the GCM don't provide suitable forecasts for areas as small as the SBS. There may be too much reliance on models in this analysis and the others done by USGS.

It is also important to know why data for 1979-2005 were not used for ice and polar bear survival and population growth relationships (i.e. why only 2001-2005?). Perhaps simply plotting an ice value (e.g., number ice free days over the continental shelf) vs. the polar bear population estimate or survival estimates would show the trend over a longer time period than 2001-2005 (i.e., 1979-2005). On page 12 it is noted that the models are calibrated with the ice free mean of 114 days for 2000-2005. Why not use the 1979-2005 mean for a better picture of the ice conditions in the SBS? Rode et al. used 1982-2006 data in their report.

It is disconcerting that (page 12) one of the models is altered because it gave unexpected results. This included an 11% positive response to bad ice years, and a negative response to good ice years giving the population a half-life of 12 days. This appears to me (and probably other non-modeler empiricists) that the models have limited utility. Testing with future observational data will tell. This is why presenting the results as hypotheses, not conclusions, is important. The claim that linking the climate and population models is a breakthrough (page 15) may be the case, but testing with future data is needed to know this.

On page 16-17 the authors argue that their results support a cause-effect relationship (between sea ice and polar bear population size or survival), while acknowledging that a cause-effect relationship can't be proven in a study such as theirs. I think that suggesting evidence for a cause-effect relationship is an exaggeration considering the analyses reported. Their results provide the grounds for a hypothesis about the future of the polar bears in the SBS, given the hypotheses about the future ice extent and duration. A statistical relationship does not prove cause-effect, and in this case there is not yet an effect for which to document a cause. They are predicting a future cause-effect relationship that needs to be tested with observational data.

The discussion about density dependence and immigration/emigration are relevant (page 17-18) and these topics may be important if this population and others begin to change as the authors predict. High fidelity to natal areas (page 18) may change as environmental conditions or population density change, and is not absolute as polar bears of both sexes have high mobility and gene flow among populations.

A reference is needed for the claim that nutritive benefits of terrestrial feeding are calorically insignificant (page 18).

This study assumes that females generally first breed at 5 years old. However, some 4-year old females probably breed in the SBS. Also, sub-adult males are identified as 2, 3, and 4 years old, but some males as young as 3 years old may breed. These observations, and inferences of reproduction by individuals and females breeding < 2.5 years apart, have been made with field and genetic data (Cronin, Amstrup, Talbot, Sage, Unpublished manuscript), and may be useful in the kinds of analyses done in this report. Also, genetic analyses (e.g., Cronin et al. 2006, Can. J. Zool. 84:655) provide estimates of emigration and immigration among populations that can be used in demographic analyses.

As in the other USGS reports, sea ice extent is defined as the area with ≥50% ice cover. Documentation that there is equal habitat value for conditions with 0%-49% ice cover is needed. Durner et al. use a 15% ice cover threshold in some analyses.

627 bears were caught in this study. The potential impact of capture and mortalities should be reported, as did Stirling et al.

Hunter et al. noted that Regehr et al. found negative effects of ice free periods on survival, breeding, and population growth. It appeared to me that Regehr et al. only found a negative relationship of survival and ice free periods > 127 days. I think the different findings of the different studies need standardization of terminology and explicit statement of results to avoid confusion or misinterpretation of results.

It is not clear to me how survival was determined by Hunter et al. or Regehr et al. It is also not clear if the identification of preferred habitat excluded consideration of other non-preferred habitat with a lower but non-zero value to polar bears.

Review of Rode et al.: Polar bears in the southern Beaufort Sea III: Stature, mass, and cub recruitment in relationship to time and sea ice extent between 1982 and 2006.
I don't know if the authors appreciate the significance of Figure 4, described on page 8. There was no significant trend in % days April-Nov in which the continental shelf was covered by ≥ 50% ice between 1982 and 2006. This means there is no trend in the ice free period length, which was the subject of the other SBS studies using only 2001-2005 ice data, and the polar bear issue in general. In other words, the longer-term ice data from 1982 to 2006 do not show a decline in the ice over time.

This seems counter to the entire argument that there is a problem. Granted the ice models predict the ice free period will increase hence-forth, but the empirical data before us don't show it for the period 1982-2006. This may mean that trends in bear measurements are not related to ice because there is no trend in the ice cover.

With no temporal trend, it seems to me that the data reflect relation-ships of bear condition and habitat conditions on *an annual basis*. In some years, there is less ice cover and therefore there is less habitat (or other environmental differences), and some measurements reflect this. I believe the lack of a trend in ice cover during the period 1982-2006 restricts the results to an annual phenomenon, not a cumulative effect. It is appropriate to hypothesize that if ice cover decreases in the future as models predict, bear condition may deteriorate, but the ice and bear data at hand don't seem to allow conclusions at this time.

The potential for differences in bear measurements to reflect annual environmental conditions may also be relevant to the assessment of stature, body mass, and BCI. Only spring measurements of stature, body mass, and BCI were used (page 5). It seems winter severity (e.g., heavy ice years have lower seal availability, see Stirling et al.) could affect these measures in the spring, and not be related to summer ice coverage.

Winter severity could also affect yearling survival and hence affect the yearling/female ratios associated with summer ice in the report. The non-significant cub/female ratio and ice relationship may reflect the mother-cub pairs not being directly subjected to winter conditions. Cub/female pairs are in dens prior to the spring ratios, and subjected only to summer conditions prior to the fall ratios so these ratios might not reflect winter conditions. If I'm not mistaken, females with year-lings will not be in dens in the winter, and survival over winter and the ensuing summer may reflect winter conditions as well as the effect of summer ice conditions. A variable of winter prey availability or winter severity may improve the biological relevance of the analysis.

A point about the ice cover data could also be clarified. Percentages (not numbers) of days of <50% ice cover are used because the number of days with ice data varied from 184-365 days per year. However, only the percentages ice free days during April to November seem to be the ones that matter for this analysis. It is not clear how this was managed. Also, if the numbers of days used for different years varied from early to mid to late summer (with increasing ice free conditions over the summer), some years may be biased toward more or less ice compared to others. For example, if one year had most of the days with data in June, it could have less ice-free days than a year with most of the days with data in

September simply because the ice progressively melts over the summer. Some standardization may help in this regard.

The point about less productivity in deeper water should address whether this includes documentation of differences in polar bear prey as opposed to productivity of another measure.

The data and analysis of the polar bear measurements (e.g. weights, lengths, etc.) by Rode et al are interesting. The data should be available so the research can be replicated.

I think there is a selective emphasis in the discussion and conclusions that the results show a negative response to ice or trend over time. For example, there is no relationship between COY/female and % days with $\geq$50% ice, and a positive relationship between yearlings/female and % days with $\geq$50% ice (Table 5). In the conclusions it is stated that "cub size and apparent survival during their first two years of life were negatively affected by years of poor ice coverage..." First, it is inappropriate to draw cause-effect inference with the use of the word "affected". The authors show a statistical relationship, not a cause-effect relationship. Second, it could also be stated that the COY/female ratios were not related to sea ice, and yearling size (more appropriately skull size and body mass) was not related to sea ice. In other words, non-significant results are still results. In this case, cubs/female didn't differ with more/less ice, but yearlings/female did, and cubs had smaller mass with less ice but yearlings didn't. These appear contradictory, but the results supporting positive relationships of ice and condition are emphasized in the conclusions.

There are a lot of good data and analyses in the report, and it is encouraging to see the use of hypotheses (page 4). However, I suggest hypotheses be stated explicitly and put into a table so the results supporting and not supporting various hypotheses are given equal validity. This would duplicate Tables 3 and 5 to some extent, but it would show consistent or inconsistent support for related hypotheses.

Example of Table with hypothesis testing results:

Hypothesis	data support hypothesis?
Negative trend of sea ice over time	no
Negative trend of skull size 1982-2006	
Adult males (older)	no
Adult males (younger)	yes
Adult females	no
Positive relationship of sea ice & COY/female	no
Positive relationship of sea ice & Yearling/female	yes
Etc.	

Review of Durner et al.: Predicting the future distribution of polar bear habitat in the polar basin from resource selection functions applied to 21st century general circulation model projections of sea ice.

In several places the report refers to loss of summer sea ice since the 1970s, and a marked decline in sea ice habitat from 1985-1995 to 1996-2006 (e.g., pages 3 and 5). On pages 13 and 19-20, declines in optimal habitat between 1985-1995 and 1996-2006 are noted in several specific areas, including the SBS. This seems to be in conflict with Fig 4 of Rode et al. who showed no trend in summer ice loss in the SBS from 1982-2006.

The analyses and results are complex, but I think Durner et al. include use of habitats with a 15% ice threshold. All the other USGS reports appear to restrict ice coverage $\geq$50% as habitat. Figure 7 of Durner et al. shows relatively high probability of selection of habitats in summer with say, 20% to 50% ice cover. The use of 50% ice cover in the other reports should be critically assessed and revised if lower ice coverage provides measurable habitat value.

Figure 7 of Durner et al. also shows considerable probability of selection of summer habitats with water depths > 1000 m. The other reports appear to use a smaller depth (e.g., 300 m) as a criterion. This should be critically reassessed in the other studies.

Density dependence and changes in ranges are discussed on page 18. Restoration of ice each winter is also discussed on page 18. These are good points to assess the long term future of the entire species given the predicted changes to ice. Reduced carrying capacity and changed distribution could be hypothesized, and tested with future observations.

APPENDIX CHAPTER 6 THE NORTH SLOPE OF ALASKA: OILFIELDS AND CARIBOU

Unpublished essay by M. A. Cronin: Federal oil for the U.S. military: A solution to the impact of sequestration on the military, enhanced national security, and economic growth. 11 September 2013.

Appendix Chapter 6 The North Slope of Alaska: Oilfields and caribou
1. **Unpublished essay by M. A. Cronin: Federal oil for the U.S. military: A solution to the impact of sequestration on the military, enhanced national security, and economic growth**

Matthew A. Cronin, Research Professor of Animal Genetics

University of Alaska Fairbanks, School of Natural Resources and Agricultural Sciences, Agricultural and Forestry Experiment Station, Palmer, Alaska

11 September 2013

I have become aware of the impact of budget cuts and sequestration on the U.S. military from recent testimony of top military leaders in Congressional hearings. It is clear that readiness, training, equipment maintenance, and most importantly the morale of our military personnel, is being compromised by funding reductions.

I am also aware of the vast natural resources under federal government jurisdiction, including oil, gas, timber, and minerals. Many of these resources are not being developed because of environmental regulations and litigation. My experience has been that natural resources can be developed without serious environmental impacts, but that regulators, scientists, and environmental groups stop development with selective use of science to predict impacts on fish and wildlife.

I submit this proposal for immediate Congressional action to help alleviate the threat to national security caused by budget reductions to the military and begin restoration of sound natural resource management on the federal estate.

The problem

Budget cuts, including sequestration, are having a negative impact on U.S. military operations and readiness. The largest costs to the military are personnel and fuel.

Relevant facts

1. The federal government's primary responsibility is national security.
2. The U.S. military is experiencing budget cuts.
3. Budget cuts decrease the military's ability to provide national security.
4. Budget cuts decrease the morale of U.S. military personnel.
5. U.S. military personnel risk their lives to protect the country.
6. The present budget cuts put national security and the safety of our forces at risk.
7. The U.S. imports oil and gas from foreign nations, some with citizens that are hostile to the U.S.
8. The federal government controls large amounts of energy resources in the western states.

9. Development of oil and gas on federal lands is prevented by excessive environmental regulation, obstruction by government agencies, and lawsuits by environmental groups.
10. Oil development on Alaska's North Slope since the 1970's has not negatively affected fish and wildlife populations. Private oil-field development has been environmentally sound, with state-of-the-art mitigation and restoration (See Figure 1).

The solution

Objective: Provide immediate funding for the U.S. military for fuel and personnel costs.

Immediately develop oil resources on federal lands with the revenue from the sale of oil going directly and exclusively to the Department of Defense (DOD) U.S. Army, U.S. Air Force, U.S. Navy, U.S. Marine Corps, and the Department of Homeland Security (DHS) U.S. Coast Guard (USCG) for:

1. Fuel costs.
2. A 10% pay raise for military personnel.

Action plan

Alaska's North Slope oil is the first operation.

Other States and other resources (e.g., natural gas, minerals, and timber) can be added to the program.

There are extensive onshore oil and gas prospects in Northern Alaska under federal government control including the Arctic National Wildlife Refuge (ANWR) and the National Petroleum Reserve Alaska (NPRA).

Congress will pass a law, with the approval of the State of Alaska legislative and executive branches, to do the following:

1. Begin immediate and extensive exploration for oil in ANWR and NPRA beginning in the winter of 2013-2014.

2. The action is for maintaining national security. Therefore, it will use existing environmental assessments and impact statements, will not require additional research, permits, or regulation, and will not allow lawsuits to prevent exploration and development. The activity will be exempted from all regulatory permits, and will comply with standard safety and environmental management practices currently in use.

3. An exploration plan developed by the State of Alaska, in partnership with the oil industry and the U.S. Geological Survey (USGS) Energy Resources Program, will be used. Private sector oil industry experts will design development operations to maximize oil production and minimize environmental impacts. North Slope residents and experienced oilfield personnel will advise with local knowledge.

4. Upon discovery and delineation of oil reserves, begin immediate leasing and rapid production, with oil being transported to the Trans-Alaska Pipeline (TAPS) for delivery to the Valdez oil terminal.

5. The above actions will occur under accelerated leasing to the private-sector oil industry.

6. Oil lease receipts and a per-barrel royalty will be allocated to the military (DOD and USCG) to be used for fuel and personnel costs. Current DOD and USCG base funding will not be reduced as a consequence of this oil revenue.

7. No federal government agency actions will be taken, or costs incurred, other than DOD and USCG direct costs to administer the project.

8. The goal is to have exploration completed no later than April 2014, with production and first oil as soon as possible in 2014-2015.

Benefits of the action

1. Enhance national security by enabling the military to train and conduct operations.

2. Enhance military personnel morale and show them that we the people care about them and appreciate their sacrifice and service.

3. Demonstrate that the USA has the resolve to maintain its military capability and develop our natural resources for our own use.

4. Accelerate USA energy independence and end dependence on foreign oil.

5. Establish the federal government as participating with, rather than hindering, the States and private sector in resource development.

6. Provide economic growth for the country and good private sector jobs for young Americans.

Example of oil revenue for DOD
Volume of oil:
bbl = barrel, Bbbl = 1 billion barrels
USGS estimates for Alaska federal onshore North Slope (Figure 2):
ANWR (10.36 Bbbl) and NPRA (0.90 Bbbl) = 11.26 Bbbl
Assume DOD uses 360,000 bbl oil/day.
11,260,000,000/360,000 = 31,278 days = 85.7 years of oil for DOD.
Oil in excess of 360,000 bbl/day can be produced to cover the costs of pay raises for military personnel.
TAPS can transport > 2 million bbl/day.

Value of oil:
Assume $100/bbl oil
11.26 Bbbl X $100/bbl = $1,126 billion = $1.126 trillion = $1,126,000,000,000 = $1.126 x 10^{12}

Assume 50% royalty for oil, so net value = $563,000,000,000 ($563 billion).

Appendix Chapter 6. The North Slope of Alaska: Oil Fields and Caribou.

Figure 1. The Central Arctic caribou herd uses habitats in and around the North Slope Alaska oil fields. Numbers of caribou in the herd are shown below (Data from Alaska Department of Fish and Game). The Trans-Alaska Pipeline (TAPS) began operation in 1977.

Central Arctic Caribou Herd Numbers

Census Year	Number of Caribou
1975	5,000
1978	5,500
1980	5,000
1981	8,537
1983	12,905
1985	15,000
1989	18,000
1991	19,046
1992	23,444
1995	18,100
1997	19,730
2000	27,128
2002	31,857
2008	66,772
2010	**70,034**

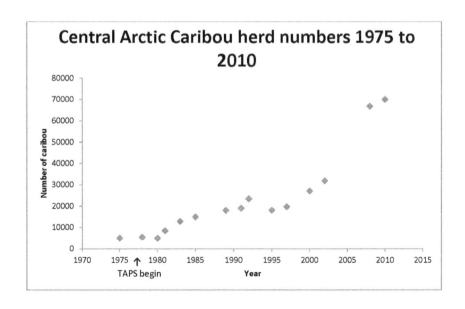

Appendix Chapter 6. The North Slope of Alaska: Oil Fields and Caribou.

Figure 2. Oil in the USA from USGS.
(http://certmapper.cr.usgs.gov/data/noga00/natl/graphic/2013/mean_conv_oil_2013_large.png)

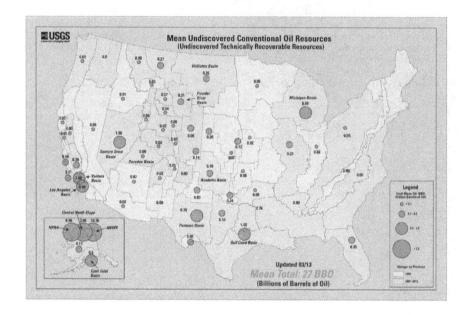

CPSIA information can be obtained
at www.ICGtesting.com
Printed in the USA
LVHW020548221019
634943LV00006B/224/P